Studies in Systems, Decision and Control

Volume 452

The series "Studies in Systems, Decision and Control" (SSDC) covers both new developments and advances, as well as the state of the art, in the various areas of broadly perceived systems, decision making and control–quickly, up to date and with a high quality. The intent is to cover the theory, applications, and perspectives on the state of the art and future developments relevant to systems, decision making, control, complex processes and related areas, as embedded in the fields of engineering, computer science, physics, economics, social and life sciences, as well as the paradigms and methodologies behind them. The series contains monographs, textbooks, lecture notes and edited volumes in systems, decision making and control spanning the areas of Cyber-Physical Systems, Autonomous Systems, Sensor Networks, Control Systems, Energy Systems, Automotive Systems, Biological Systems, Vehicular Networking and Connected Vehicles, Aerospace Systems, Automation, Manufacturing, Smart Grids, Nonlinear Systems, Power Systems, Robotics, Social Systems, Economic Systems and other. Of particular value to both the contributors and the readership are the short publication timeframe and the world-wide distribution and exposure which enable both a wide and rapid dissemination of research output.

Indexed by SCOPUS, DBLP, WTI Frankfurt eG, zbMATH, SCImago.

All books published in the series are submitted for consideration in Web of Science.

Jagannath Singh · Debasish Das · Lov Kumar ·
Aneesh Krishna
Editors

Mobile Application Development: Practice and Experience

12th Industry Symposium in Conjunction with 18th ICDCIT 2022

Springer

Editors
Jagannath Singh
School of Computer Engineering
Kalinga Institute of Industrial Technology
Bhubaneswar, India

Debasish Das
Enterprise Architect Cyber Security Practice
Tata Consultancy Services
Bhubaneswar, India

Lov Kumar
Department of Computer Science and Information Systems
Birla Institute of Technology and Science
Hyderabad, India

Aneesh Krishna
School of Electrical Engineering, Computing and Mathematical Sciences
Curtin University
Perth, WA, Australia

ISSN 2198-4182 ISSN 2198-4190 (electronic)
Studies in Systems, Decision and Control
ISBN 978-981-19-6892-1 ISBN 978-981-19-6893-8 (eBook)
https://doi.org/10.1007/978-981-19-6893-8

This Springer imprint is published by the registered company Springer Nature Singapore Pte Ltd.
The registered company address is: 152 Beach Road, #21-01/04 Gateway East, Singapore 189721, Singapore

Preface

In last decade, there is a huge gain in the mobile application usages and development. Due to the portability and convenience of mobile devices, mobile apps have surpassed traditional desktop applications as the primary way of accessing the Internet. Many users depend on their smartphones for daily tasks such as reading, shopping, paying and chatting through mobile apps. Not only the mobile application has an impact for user but also it plays an important role in business. Many business companies are earning revenue using mobile application. So, the mobile application has an impact on society. Different from traditional desktop applications, mobile apps are typically developed under time-to-market pressure. The process of mobile app development is still not standardized, and the methodology used for the development of mobile app is still inadequate. There is still lack of research works and insufficient understanding of real issues and challenges faced in the mobile app development.

This book focuses on software engineering related research and practice supporting any aspects of the mobile app development. It contains the chapters that address requirements, analysis, implementation, maintenance, evolution, testing, security or any other aspects of mobile apps development. We can observe the significant use of mobile technology in fields like business, agriculture, production, banking and many more. So, this book is cantering the detail study of these fields, application of mobile technology into it and challenges.

While developing the mobile applications for users, developers have to choose appropriate application development model and develop high-quality apps on time with less budget. For this, knowledge of app development model plays an important role. The mobile applications are different from web applications in the amount of resource utilization, because the number of resources available in mobile phones will be less compared to computers. In the first Chapter "A Survey on Mobile Application Development Models", a review is presented to find, analyze, compare and contrast various mobile app development models and the software development standards needs to be followed while developing mobile applications.

There are many fields where mobile technologies are becoming very popular. Smart farming is one of them where mobile technologies are used in mapping of soil, fertilizer optimization, climatic conditions and finally in maximizing the quality and quantity of agricultural produce. Furthermore, mobile digital technologies have enabled farmers to collect critical information on soil quality, such as nutrient levels, water levels, pests and disease spread, via various sensors, satellites and drones, among other things. Chapter "Mobile Technology for Smart Agriculture: Deployment Case for Pearl Millet Disease Detection" highlights the importance of mobile technology for farmers and deployment of app for disease detection in pearl millet crop.

In the thrid Chapter "Energy Consumption-Based Profiling of Android Apps", one technique to test the power efficiency of any android application is presented. This is done by checking the complexity of the program by counting the number of cycles or loops present in the code of the android application and calculating the energy consumption through a formula. It will help users to decide which application is more efficient and consumes less power so that they can maximize their usage by choosing the app which consumes less power and save their mobile phones from draining their battery very fast.

In the fourth Chapter "Impact of COVID-19 on IT Business", the authors have presented the study of impact of COVID-19 on the IT industry by doing empirical research on a variety of issues that have influenced current Indian IT enterprises. As COVID-19 has had a significant influence on many firms, this study will help business owners to identify the causes that are causing them to close their doors due to financial losses. Specifically, machine learning algorithms will aid in the classification of data into several categories in which organizations operate, and companies will be able to forecast whether they will profit or suffer a drop in income based on this information.

The security of smartphones becomes a crucial factor, especially due to malware. Chapter "ANNDroid: A Framework for Android Malware Detection Using Feature Selection Techniques and Machine Learning Algorithms" presents malware detection frameworks based on the principle of simple neural network and regression analysis. Proposed machine learning techniques are applied on five million distinct android apps. In addition to that, this chapter also paid attention toward feature selection techniques such as rough set analysis (RSA) and principal component analysis (PCA) when they are implemented for malware detection.

In today's environment, a lot of emphasis is given on the global adoption of standard practices for mobile application development. In the sixth Chapter "DanVeer: A Secure Resource Funding Mobile Application", the standard practices for mobile application development are reviewed and a new app 'DanVeer' has been proposed which is a crowdfunding application. It is a native android application designed and developed in Java and implements concepts of blockchain. The chapter focuses on the various software engineering aspects of mobile application development such as architecture, testing and debugging of apps, app review and analysis, applications beyond smartphones and tablets, maintenance of apps, app security and privacy and similar others.

In the seventh Chapter "Mobile Data Analytics: A Comprehensive Case Study", the authors have presented the concepts of mobile data analytics. Mobile data analytics has been considered to be a field of enthusiastic development among the IT experts and business executives. Mobile data analytics is able to manage big data analytics on resource constricted devices. It has been found that most of the companies have found to improve their customer experience and provide the users with a better user experience after executing mobile data analytics of the data that they had collected from them.

When refactoring a program or piece of software, the goal is to make it more efficient without changing its usefulness. A large-scale experiential study that examines the evolution, introduction and refactoring operations for the enhancement of code quality is presented in the eighth Chapter "Method Level Refactoring Prediction by Weighted-SVM Machine Learning Classifier". Five open source projects have been considered by the authors to apply weighted SVM and SVM with SMOTE as machine learning classifier for refactoring prediction.

Currently, there exists very few applications for fishermen's protection and ease of fishing. Fishermen encounter challenges such as trespassing past country borders by mistake since they do not know where the border is, causing problems with their citizenship and maybe being proclaimed criminals, and also fishermen are unable to locate natural resources (e.g., fishes). In the ninth Chapter "OCEANDROID", the authors proposed an android app named OCIANDROID that addresses these issues. In addition, the density of fish in a specific location will be measured and reported, allowing fishermen to fish more readily.

In the tenth Chapter "Innovation Propensity of Firms and the Interplay of Institutional Ecosystem—A Longitudinal Study from G-20 Middle-Income Countries" analyzes the impact of a unique set of institutional ecosystem factors on the innovation propensity of firms in some of the most important middle-income countries of the world. As these middle-income countries embrace innovation as growth levers, it is pertinent to analyze the impact of emerging institutional policies on innovation strategies adopted by firms in these countries. In this study, the authors have focused on the G20 block of nations, and among the G20, we further focus our attention on all the permanent members which are middle-income economies.

Creating multiple fake accounts to reap the benefits of an online service provider has always been an issue for corporations. The main objective of eleventh Chapter "Unique and Secure Account Management System Using CNN and Blockchain Technology" is to suggest an account management service that limits each user to have only one account that is private and secure in every aspect. The proposed model also allows the users to have a unique digital account without sharing personal details with the service-providing companies. It uses facial recognition algorithms for unique identification purposes and blockchain technology for immutable and secure data storage to implement this idea.

In the twelfth Chapter "Model for Mobile App-Based Premium Calculation for Usage-Based Insurance (UBI) of Vehicles" contains the design of a mobile application that helps in calculating the usage of a vehicle using GPS tracking devices attached with the vehicle, which then being used to generate the risk and usage factors of the vehicle. It will be also utilized by the concerned agencies to calculate the next premium of an insurance. Enough information is being generated and stored that are being used by the model for the further processing.

Accurate estimation of attributes such as effort, quality and risk is of major concern in software life cycle. In the thirteenth Chapter "Web Service Anti-patterns Detection Using CNN with Varying Sequence Padding Size", Chidamber and Kemerer software metrics suite has been considered to provide requisite input data to train the artificial intelligence models. Two artificial intelligence (AI) techniques have been used for predicting maintainability, viz. neural network and neuro-genetic algorithm (a hybrid approach of neural network and genetic algorithm).

Topics presented in each chapter of this book are unique to this book and are based on unpublished work of contributed authors. In editing this book, we attempted to bring into the discussion of all the new trends and experiments that have made on mobile application development. We believe this book is ready-to-serve as a reference for a larger audience such as system architects, practitioners, developers and researchers.

Bhubaneswar, India — Jagannath Singh
Bhubaneswar, India — Debasish Das
Hyderabad, India — Lov Kumar
Perth, Australia — Aneesh Krishna
January 2022

Acknowledgements

First and foremost, praises and thanks to God, the Almighty, for His showers of blessings throughout our journey to complete the edited book successfully.

We would like to express our heartfelt gratitude to our families, friends, colleagues and well-wishers for their constant support throughout the preparation of this book. We gratefully thank all the contributors, who allowed us to quote their remarks and work in this book. In particular, we would like to acknowledge their hard work and cooperation during the revisions of their chapters.

We would also like to acknowledge the valuable comments of the reviewers which have enabled us to select these chapters out of so many chapters received. It helped us to improve the quality of the chapters.

We express our special thanks to acknowledging and appreciating Springer team for their continuous support throughout the entire process of publication.

Finally, we are extending our thanks to the researchers, who motivated us, and hope this work will certainly inspire and guide them to reach their destinations. We also thank KIIT DU authorities for having and permitting us to proceed with this book.

Bhubaneswar, India — Jagannath Singh
Bhubaneswar, India — Debasish Das
Hyderabad, India — Lov Kumar
Perth, Australia — Aneesh Krishna
January 2022

Contents

About the Editors

Jagannath Singh is an Assistant Professor in the School of Computer Engineering at Kalinga Institute of Industrial Technology, Bhubaneswar, India. He received his Ph.D. and M.Tech. degrees from the National Institute of Technology (NIT) Rourkela, India, in 2016 and 2012, respectively. He has published several research articles in reputed conferences and journals. His research interest includes software engineering, artificial neural networks, and machine learning. He is guiding three full-time Ph.D. students.

Debasish Das has over 17 years of experience in product engineering, software systems, cyber security, industrial IoT and quality management focusing on research and innovation, strategy, leadership and management. He has worked in interdisciplinary fields of design, architecture, performance engineering, application security, QA and automation along with business development and consulting. He's a founding member of TCS Video Engineering and Analytics Lab. As an enterprise architect, he is currently working as a core member of TCS Enterprise Vulnerability Management CoE. Debasish completed his M.Tech. from BITS Pilani. His research area focuses on emerging cyber security paradigms e.g. malware resiliency, cyber risk, and intelligent vulnerability remediation systems. He also focuses on video engineering R&D for next-gen ecosystems using 5G, edge computing, and industrial IoT. Over the years, he has filed multiple patents, developed innovative platforms, organized tech events, and won multiple innovation awards for his research works which have been published in journals of national and international repute.

Lov Kumar is an assistant professor in the Department of Computer Science and Information Systems, BITS Pilani, Hyderabad. He received his Ph.D. in Computer Science and Engineering from the National Institute of Technology (NIT) Rourkela. His current research interests are in the area of mining software repositories, machine learning, text analysis, testing of AI systems, software analytics, and social media analytics. He has delivered over 25 invited talks and published over 50 refereed publications in international conferences and journals, and two book chapters.

Aneesh Krishna is an associate professor with the School of Electrical Engineering, Computing, and Mathematical Sciences, Curtin University, Australia. He holds a Ph.D. in computer science from the University of Wollongong, Australia. He was a lecturer in software engineering at the School of Computer Science and Software Engineering, University of Wollongong, Australia. His research interests include AI for software engineering, model-driven development/evolution, requirements engineering, agent systems, formal methods, data mining, computer vision, machine learning, bioinformatics, and renewable energy systems. He has published more than 130 articles in reputed journals and international conferences. His research is (or has been) funded by the Australian Research Council (ARC), and various Australian government agencies (like NSW State Emergency Service) as well as companies such as Woodside Energy, Amristar Solutions, and Autism West Incorporated.

A Survey on Mobile Application Development Models

A. N. Shwetha, R. Sumathi, and C. P. Prabodh

1 Introduction

Increase in the number of mobile application users has made the mobile application development field a promising one. There are typical challenges which need to be addressed while developing mobile applications due to less time to market, different development models and rapidly evolving technology. The first phase in mobile application development is to identify end users, identify the competition exist in the market for similar apps, specify objectives of application and selection of suitable model for application development. The application development model consists of sub-models like the information model, the process model and GUI model. The information model consists of object-oriented elements like classes, associations, aggregations, etc. It is not only used to generate underlying information but also used to generate UI design with concerned input and output. The graphical user interface is nothing but front end of application which defines look and feel of app.

The GUI model consists of elements like buttons, menus, pictures, pages, text fields, radio buttons, etc. The page in the application can be developed to specify the purpose of page like editing data, viewing underlying data, etc. The process model has the information regarding which software development process needs to be considered like waterfall model, agile model, spiral model, etc. Depending on

A. N. Shwetha (✉) · R. Sumathi · C. P. Prabodh
Department of CSE, Siddaganga Institute of Technology, Tumakuru, Karnataka, India
e-mail: shwethaan@sit.ac.in

R. Sumathi
e-mail: rsumathi@sit.ac.in

C. P. Prabodh
e-mail: prabodh@sit.ac.in

J. Singh et al. (eds.), *Mobile Application Development: Practice and Experience*,
Studies in Systems, Decision and Control 452,
https://doi.org/10.1007/978-981-19-6893-8_1

the requirements identified for a particular application and the purpose of mobile application, a specific software development methodology can be adapted. Each software development methodology has its own advantages and advantages, and they suit particular type of mobile application.

The model which suits for traditional software or web app does not suit for mobile applications because they have their own pros and cons.

Mobile Applications

Pros:

1. These are faster than web apps
2. Provides more functionality
3. No need of internet connectivity
4. Provides security
5. Easy to implement.

Cons:

1. Increased development costs
2. No compatibility between different platforms
3. Maintenance is expensive
4. Difficult to get approval from app store.

Web Applications

Pros:

1. No installation
2. Maintenance is easy
3. Self-upgradation
4. Quick implementation
5. Does not require approval from app store.

Cons:

1. Need internet connectivity
2. Slow
3. Less availability
4. Less guaranteed security.

2 Literature Survey

Kaur and Kaur [1] have done a detailed review to find and compare various techniques of test estimation for mobile applications compared to traditional applications. Also, the differentiation is made between traditional software and mobile applications. The review is done by data extraction, by formulating research questions and by detailed selection studies. Vaupel et al. [2] presented a language that can be used to model mobile applications and an infrastructure for android apps which supports different variants. The model-driven development is demonstrated using two applications such as phone manager to maintain contacts and a conference guide to organize conference by organizers and for participants.

Seiren et al. [3] done a comparative study on the approaches adapted for current mobile app development. Also, a detailed comparison is made between existing approaches which helps developer to select particular approach depending on the requirements of software. Hanif et al. [4] have done a comprehensive literature review to understand the insights into mobile application development by considering various Scopus journal papers in the domain of android development and mobile application development. Balagtas-Fernandez and Hussmann [5] proposed a mobile application specific graphical language for model-driven development. Also, a tool to generate code automatically from graphical model is also invented.

Masi et al. [6] done a research review to help mobile application developers to select a technology which suits the requirements by giving practical guidelines. The different research methods used by authors are survey, interview and case study. The review results in a set of available models, experts experience and advanced platforms. Jayatilleke et al. [7] conducted a review on mobile application development by sessions conducted in various institutes. Khandelwal and Tyagi [8] done a review to identify the suitability of particular process model for the life cycle of mobile application development. Yu [9] conducted a review to compare and contrast various software development models available and also identified the most commonly used development models because of their advantages. Ma et al. [10] discussed about various kinds of mobile application can be developed using android platform.

3 Mobile Application Development Models

There are five different mobile application development models which exist. They are

1. Native application development
2. Hybrid application development
3. Cross-platform application development
4. PWA development
5. Desktop application development.

Native application development model is used to create mobile app which supports single platform/operating system. It uses programming language which is the operating system dependent. For example, Java or Kotlin for android OS and swift or objective C for IOS. Native application development model will have access to all features in a feature set of device, so that it gives a good performance. Tools that can be used in native application development model are android studio, android IDE and ATOM.

Advantages:

1. Gives best performance
2. Provides more security
3. Will be more interactive
4. Full feature set of devices can be accessed
5. Fewer bugs during development.

Hybrid application development model combines elements of web applications with mobile application. First, codebase needs to be developed using web technologies like HTML and CSS. Then, the codebase will be wrapped inside the container. The container is web view which will act as a browser to load the application. Native plugins specific to a mobile device can be installed specific to mobile device, so that features of mobile device can be accessed through web application. Tools to be used in hybrid app development are Cordova, Ionic, etc. Example applications developed using hybrid application development model are Gmail, Evernote, etc.

Advantages:

1. Uses agile process model
2. Easy to find resources
3. Code reusability
4. Less development time and cost.

Cross-platform application development model makes use of native rendering engine. Bridges are used to connect codebase written in framework-dependent programming language to native components. Cross-platform applications do not depending on platforms. They are easy to implement, cost-effective and provide good functionality. Tools to be used in cross-platform application development model are React Native, NativeScripts and Flutter. Examples of applications developed using cross-platform application development model are Instagram, GoogleAds.

Advantages:

1. Quick time to market
2. Single source code
3. Easy implementation
4. Easy maintenance
5. Same application can run on different devices.

Progressive web application development is an alternative approach to traditional mobile application development. Here, the application need not to be installed so that storage is not required. These are the web applications which utilize the capabilities of a browser to provide mobile application like user experience. Tools to be used in progressive web application development model are React, Angular JS. Examples of applications developed using progressive web application development model are e-commerce applications, Flipkart.

Advantages:

1. Easy to develop
2. Easy to distribute and update
3. Can run on any device
4. Easy to access.

Desktop application development model is the most widely used model nowadays to develop cross-platform desktop applications. Electron can be used as a tool to develop desktop applications. Some of the desktop applications are Chime, slack.

Advantages:

1. High efficiency
2. Easily scalable
3. Security
4. Easy to maintain.

4 Comparison of Different Development Models

First, three development models discussed above are considered for comparison. The comparison of models gives us a clarity on which development model can be used to develop a particular application. The comparison of three models is done here by considering both technical aspects and non-technical aspects. Table 1 gives the comparison of different models based on various technical aspects like performance, native feature access, reusability, learn once-write anywhere, UI components and availability of third-party libraries.

Table 1 Comparison of different models on technical aspects

	Native application development model	Hybrid application development model	Cross-platform application development model
Performance	More	Less	More
Native feature access	Direct access	Uses third-party plugins	Uses-third party plugins and inbuilt APIs
Reusability	Less	More	More
Learn once, write anywhere	No	Yes	Yes
UI components	Has built-in UI components	Has built-in UI components	Uses third-party plugins + third-party plugins
Third-party libraries	Available	Available	Available

Figure 1 depicts the flowchart for identification of development model based on requirements identified for applications. Initially, some of the parameters need to be evaluated to find the model which suits the particular mobile application. If the development time is less, performance is not playing major role, application should be released as early as possible to market and hardware functionalities of mobile like camera, memory are not required, then hybrid app development can be considered for application development. If hardware functionality of mobile is required and there is less time to release application to market, then cross-platform development model suits the application. If performance of app is the main criteria and multiple teams can work with different platforms for the same application, then native app development can be considered for development of application.

Table 2 gives the comparison of different models based on various non-technical aspects like community support, time to market, development cost, hiring, development time and maturity.

5 Software Development Methodologies

There are different software development methodologies which can be adapted while developing mobile applications. Depending on the nature and clarity on app requirements, a particular development methodology can be adapted. Different software development methodologies available are:

1. Waterfall methodology

   This methodology is the simplest one compared to other methodologies. The requirements should be clear, unambiguous and simple to select. Here each phase needs to be completed before moving onto the next phase. In waterfall methodology, there is no provision to move back to previous stages.
2. Prototype methodology

   Prototype model is improved version of waterfall model. Here, initially the prototype will be developed to understand requirements clearly. Feedback will be taken from each stakeholder to verify the correctness. This model reduces the failure of application.
3. Spiral methodology

   It is a risk-driven model. The requirements of project are divided into multiple groups. Each group of requirements is implemented in a separate iteration. Usually, the complex requirements are implemented in first iteration to mitigate risks. It is suitable for complex projects.
4. Agile methodology

   It is also an iterative model. This model is used when the software requirements are dynamic and evolving. The app will be implemented in multiple cycles where the output of each model will contribute to the end result of software. This is the well-known model compared to others.

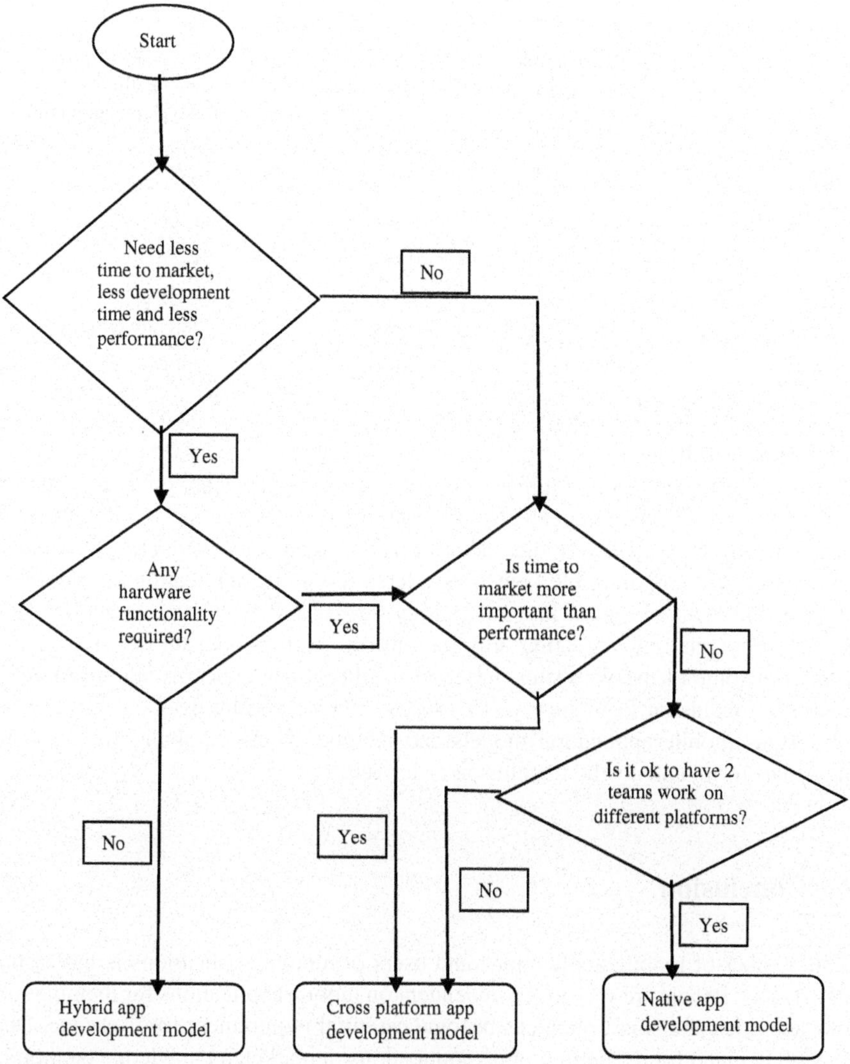

Fig. 1 Flowchart for selecting development model

5. Lean startup methodology

   The main intention behind this model is to offer software solution to startups. This model makes the developer to learn different aspects of development from the app which is developed. It focuses on fast and low-cost applications.

Table 2 Comparison of different models on non-technical aspects

	Native application development model	Hybrid application development model	Cross-platform application development model
Community support	More	Less	More
Time to market	Takes more time	Takes less time	Takes less time
Development costs	High	Less	High
Hiring	More	Less	Less
Maintenance costs	More	Less	Less
Development time	More	Less	Slightly more
Maturity	More	More	More

Threat to Validity

There will be three types of threats for research survey. They are construct validity threat, internal validity threat and external validity threat. Construct validity threat is failure to cover all relevant studies. This threat is mitigated by conducting an exhaustive search by using multiple keywords related to mobile application development models and methodologies. Internal validity threat deals with data extraction. This threat is mitigated by extracting information from various relevant research papers and from some of the Web sites. External validity threat concerns about failure to derive a conclusion from survey. This threat is mitigated by detailed comparison of various mobile application models and methodologies in tabular form and in flowchart form, and a valid conclusion is derived.

6 Conclusion

The number of mobile applications and users of mobile applications is increasing day by day. People are very much dependent on mobile applications for their day-to-day activities and smart phone as become essential requirement for everyone. Due to high demand for different mobile applications in market, a knowledge on mobile application development is very much essential and a requirement for developers. So, in this paper, a detailed analysis of different models available for mobile application development is provided along with its advantages. A comparison is given for development models by considering various technical and non-technical aspects, which will help the developer to select particular mobile app development based on application requirements. Various software development methodologies are also suggested for different kinds of mobile applications.

References

1. Kaur, A., Kaur, K.: Systematic literature review of mobile application development and testing effort estimation. J. King Saud Univ. Comput. Inf. Sci. (2018)
2. Vaupel, S., Taentzer, G., Harries, J.P., et al.: Model-Driven Development of Mobile Applications Allowing Role-Driven Variants. Springer International Publishing, Switzerland (2014)
3. Al-Ratrout, S., Tarawneh, O.H., Altarawneh, M.H., et al.: Mobile application development methodologies adopted in Omani market: a comparative study. Int. J. Softw. Eng. Appl. (IJSEA) **10**(2) (2019)
4. Hanif, S.J., Drave, V.A., Bhatt, P.C.: Mobile application development: a comprehensive and systematic literature review. In: International Conference on Industrial Engineering and Operations Management (2019)
5. Balagtas-Fernandez, F.T., Hussmann, H.: Model-driven development of mobile applications. In: 23rd IEEE/ACM International Conference on Automated Software Engineering (2008)
6. Masi, E., Cantone, G., Calavaro, G.: Mobile apps development: a framework for technology decision making. Soc. Inf. Telecommun. Eng. (2013)
7. Jayatilleke, B.G., Ranawaka, G.R., Wijesekera, C., Kumarasinha, M.C.B.: Development of mobile application through design-based research. Open Univ. J. (2018)
8. Khandelwal, A., Tyagi, G.: Review paper on suitability of traditional prototype model and spiral model used for mobile application development life cycle. Int. J. Eng. Res. Technol. (2015)
9. Yu, J.: Research process on software development model. IOP Conf. Ser. Mater. Sci. Eng. (2018)
10. Ma, L., Gu, L., Wang, J.: Research and development of mobile application for android platform. MUE 30 (2014)

Mobile Technology for Smart Agriculture: Deployment Case for Pearl Millet Disease Detection

J. Pramitha, Vijila Gnanaraj, X. Anitha Mary, D. Joel Prasanth, and I. Johnson

1 Introduction

The agriculture industry is experiencing a transition that appears to be very promising, as it will enable this main sector to reach new levels of farm productivity and profitability [1, 2]. Throughout human history, ensuring food security has been a global priority, and the global food crisis of 2007–2008 highlighted the significance of raising both quantity and quality of food production [3]. Precision agriculture, which entails applying inputs when and where they're needed, has emerged as the third wave of the contemporary agricultural revolution, and it is now being bolstered by an increase in farm knowledge systems due to the increased availability of data. On addition, when it comes to the environment, new technologies are increasingly being used in farms to ensure the long-term viability of agricultural produce. Smartphones have a considerable market share among various user segments due to their usefulness, ease-of-use and affordability among the technologies produced in the last few decades [4]. The number of people who own a smartphone is increasing. The number of users is expected to exceed 2 billion by 2016 [5].

J. Pramitha · V. Gnanaraj · X. Anitha Mary (✉) · D. J. Prasanth
Department of Robotics Engineering, Karunya Institute of Technology and Sciences, Coimbatore, Tamil Nadu, India
e-mail: anithajohnson2003@gmail.com; anithamary@karunya.edu

J. Pramitha
e-mail: pramithaj@karunya.edu.in

I. Johnson
Department of Plant Pathology, Tamil Nadu Agriculture University, Coimbatore, Tamil Nadu, India

J. Singh et al. (eds.), *Mobile Application Development: Practice and Experience*, Studies in Systems, Decision and Control 452,
https://doi.org/10.1007/978-981-19-6893-8_2

2 Importance of Computer Technology in Agriculture

Farmers benefit greatly from ICT as a decision support system. Farmers may get up-to-date information about agriculture, weather, new crop types and novel strategies to boost productivity and quality control via ICT. Farmers can use information and communication technologies to transmit precise and accurate information at the proper time, allowing them to benefit from it. Farmers can design types of crops, implement proper agricultural methods for cultivating, harvesting, post-harvesting and marketing their produce using the decision support system provided by ICT. In agriculture, several sorts of information are needed depending on the agro-climatic zones, the size of land holdings, the types of crops farmed, the technology used, market orientation, weather conditions and so on. According to several researchers, the majority of farmers consider the "question and answer service" to be the ideal facility for getting individualized solutions to their specific problems. Several app have been developed worldwide [6–11]

3 Different Mobile App for Smart Farming in India

A farmer's best buddy in farming can be a farming app that increases productivity without costing a thing. It is available for free download from the Google Play store.

3.1 Digital India Program

Digital India, an initiative established by Indian Prime Minister Narendra Modi in 2015 to encourage digital literacy and the development of digital infrastructure, looks to be supporting Rural India in achieving agricultural success.

3.2 Kisan Suvidha

The app is easy to use and includes information on current weather conditions as well as forecasts the information for the next days, market pricing of commodities/crops in the nearby town, fertilizers, seeds, machinery and so on. The app's popularity is widened by the fact that it may be used in many languages.

3.3 IFFCO Kisan Agriculture

Its purpose is to help Indian farmers make better decisions by giving them with customized information tailored to their unique needs. The user can also access a range of informational modules throughout the profiling stage, such as agricultural advice, weather, market pricing in the form of text, photography, audio and videos in their preferred language. The app also includes contact information for Kisan Call Center Services, including phone numbers.

3.4 RML Farmer-Krishi Mitr

It is a valuable farming software that keeps farmers up to date on commodities and mandi pricing, pesticide and fertilizer usage, farm and farmer news, weather forecasts and advisory. It also offers agricultural advice and information on the government's agriculture policies and programs. Users can choose from around 450 crop types, 1300 mandis and 3500 weather locations spread across 50,000 communities across 17 Indian states, according to the official. It is equipped with capabilities that allow it to assess or offer information on many elements of farming practices.

3.5 Pusa Krishi

It is a government app that was created in 2016 by the Union Agriculture Minister with the purpose of supporting farmers in accessing information on technology produced by the Indian Agriculture Research Institute (IARI) that will help farmers increase their returns. The app also informs farmers about new crop varieties developed by the Indian Council of Agriculture Research (ICAR), resource-saving cultivation practices and farm machinery, and its implementation would help farmers boost their returns.

3.6 Agri App

It is a farmer-friendly software that gives you all you need to know about crop productivity, crop protection and other agriculture-related services. Chatting with experts, video-based learning, the most recent news and online markets for fertilizers, insecticides and other products are all included in this app.

3.7 Crop Insurance

It is an excellent software that helps farmers calculate insurance premiums for alerted crops as well as provide information on cut-off dates and company contacts for their crop and region. For farmers, it functions as a reminder and an insurance calculator. It can also be used to get information on any notified crop in any notified area's standard sum insured, extended sum insured, premium details and subsidy information. It also has a connection to its website, which is designed to serve all stakeholders.

3.8 Agri Market

The app's objective is to keep farmers informed about crop prices and discourage them from participating in distress sales. Farmers may get information on crop pricing at markets within 50 km of their device's location with the AgriMarket Mobile App.

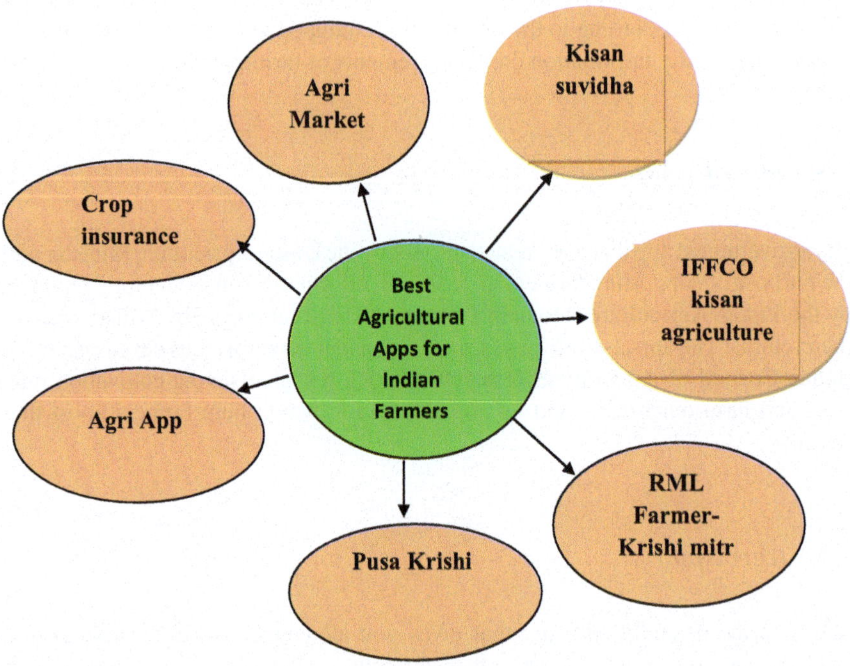

4 Structure of Android APP Deployment

Activities, services, broadcast receivers and content providers are just few of the components that make up an Android application.

- An activity is a screen containing the user interface.
- A service is a component that performs long-running processes in the background. For example, while the user is working on another program, the service can play music in the background.
- Other programs or the system send broadcast messages to broadcast receivers, which they respond to. For example, an application may notify another program that data has been downloaded and is ready to use.
- Content providers supply data from one application to another on request. Such requests are handled by the methods of the content resolver class.

5 Mobile Deployment for Leaf Disease Detection in Pearl Millet Crop

Android is a mobile platform has become easier because of the application framework. For example, if the user will need Wi-Fi to use the application, you can use the Wi-Fi manager, and if the corresponding methods are used in the program, it can easily be implemented in the program.

5.1 Android Project Structure

When a new project is created, many other folders are created inside the project. The folders created are src, gen bin, libs, assets and manifest. The resource files created are layout, drawable, menu and values. For any single screen in android, two files are required, Java and XML file. The Java file is the activity and the XML file is the layout. For instance, the XML file helps us create the user interface like the buttons, and once the button is clicked, the action to be performed is written in the Java file.

In this application "pearl millet," the details about the various diseases have been shared. Knowledge about diseases is vital as it plays an important role in getting a good yield. The opening page consists of the image and the name of the crop that has been discussed in the mobile application (Fig. 1).

An activity page is designed using various layouts such as relative layout, linear layout, constraint layout, scroll view, grid view, list view, etc.

In relative layout, the position of the elements can be specified in relation to the other elements in the activity. In linear layout, the elements are arranged in a linear fashion. There are two orientations—horizontal (elements are arranged horizontally one after the other) and vertical (elements are arranged vertically). Constraint layout

Fig. 1 Front view

is used for building large and complex activities. Constraint layout is flexible to use. Scroll view—when large data has to be displayed in the activity, the data can be scrolled and viewed which will avoid overlapping of text in the activity. Thus to achieve a screen where the information can be scrolled, a scroll view is required. In grid view, the data is displayed in the form of two-dimensional grids (rows and columns). List view is similar to the grid view, except that they do not have rows and columns but the data is displayed as a list one after the other. In this application, we have used the grid view along (Fig. 2) with a relative layout to display our data (Fig. 3).

The manifest file gives the complete description of the project such as the package name, application name and icon, number of activities and services. The Android platform will read this file before reading the application. To use certain additional features, the dependencies can be added into the Gradle so that they can be used in the program. Gradle is a build tool that can be used to build and manage processes. In this application, the CardView dependency has been added (Fig. 4).

To display the text in the activity, the user interface element TextView is used. To make the text editable by the users or in cases where the user has to enter their details, the user interface element EditText is used. The ImageView is used to display images. The elements can be dragged and dropped into the workspace. The elements can also be adjusted using codes. Once the elements are placed in the XML file, the

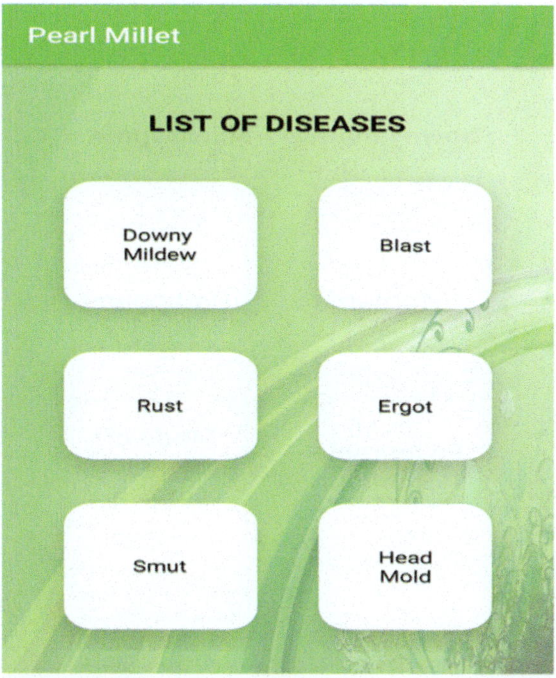

Fig. 2 Grid view with relative layout for various disease

processes to be performed are written in the Java class file. So to get the resource file into the java file R.Java class is used. Thus, we specify an ID for each element and call them in the Java code (Fig. 5).

To invoke the next activity when an option is selected, we use intent. It is a message object that invokes an activity. To start another activity in the application, the intent is built and the startActivity() method is called to send the intent to the intended activity.

6 Conclusion

ICTs have the potential to improve agriculture in underdeveloped countries by increasing access to markets and information for issue solving. Incorporating ICT opportunities for the next generation of millions of smallholder farmers could also help to close the urban–rural gap. Due to restrictions in existing delivery methods, the majority of agriculture-related *m*-services now offered in the developing world only provide basic functionality. Mobile technology is rapidly changing, and this may alter in the near future.

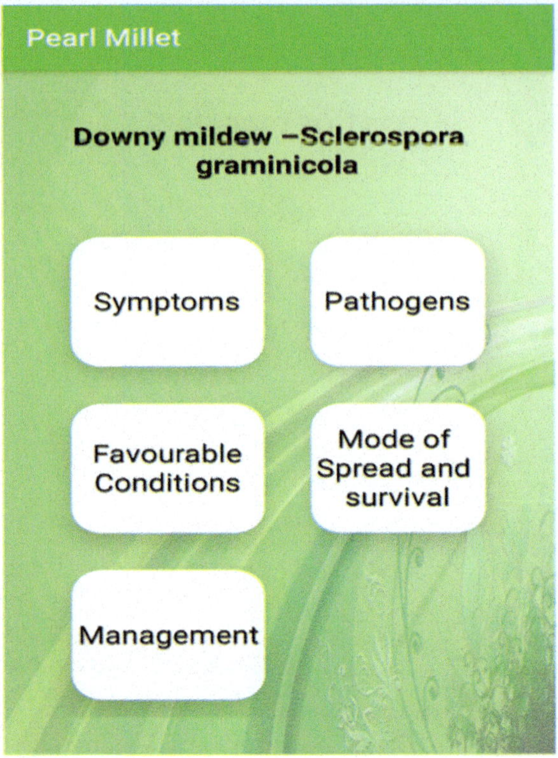

Fig. 3 CardView

```
dependencies {
    implementation fileTree(dir: 'libs', include: ['*.jar'])

    implementation 'androidx.appcompat:appcompat:1.3.0'
    implementation 'androidx.constraintlayout:constraintlayout:2.0.4'
    testImplementation 'junit:junit:4.12'
    androidTestImplementation 'androidx.test.ext:junit:1.1.3'
    androidTestImplementation 'androidx.test.espresso:espresso-core:3.4.0'
    implementation 'androidx.cardview:cardview:1.0.0'
}
```

Fig. 4 Pseudo for CardView dependency

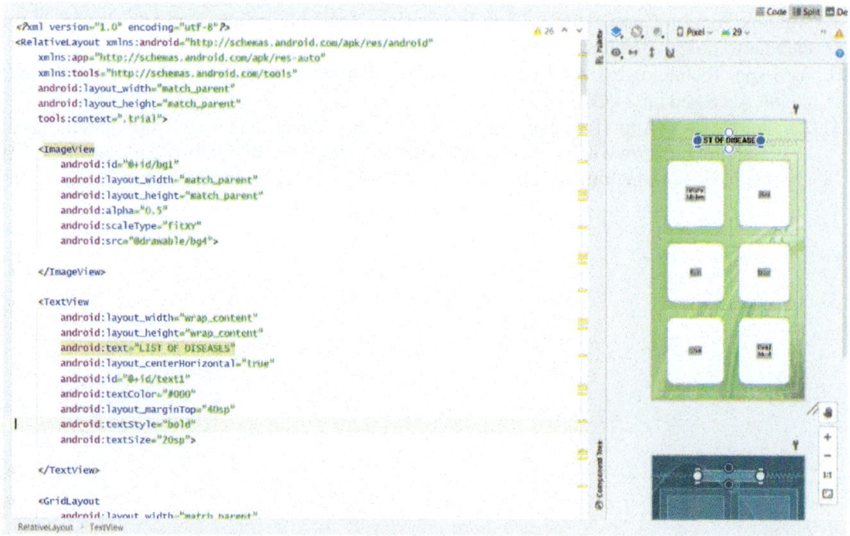

Fig. 5 XML code for the page with the list of diseases

References

1. Folnovic, T.: Smart Agriculture on Smartphones. https://blog.agrivi.com/post/smart-agriculture-on-smartphones (2021)
2. Himesh, S., Rao, P., Gouda, K.C., et al.: Digital revolution and big data: a new revolution in agriculture. CAB Rev. **13**, 1–7 (2018)
3. Sasson, A.: Food security for Africa: an urgent global challenge. Agric. Food Secur. **1**, 2 (2012). https://doi.org/10.1186/2048-7010-1-2
4. Henze, J., Ulrichs, C.: The potential and limitations of mobile-learning and other services in the agriculture sector of Kenya using phone application. In: 12th European International Farming Systems Association (IFSA) Symposium, Social and Technological Transformation of Farming Systems: Diverging and Converging Pathways, 12–15 July 2016, pp. 1–11. Harper Adams University, Newport, Shropshire, UK (2016)
5. Clark, D.: 2 Billion Consumers Worldwide to Get Smart (Phones) by 2016. http://www.emarketer.com/Article/2-Billion-Consumers-Worldwide-Smartphones-by-2016/1011694 (2014)
6. Mendes, J., Pinho, T.M., Neves dos Santos, F., Sousa, J.J., Peres, E., Boaventura-Cunha, J., Cunha, M., Morais, R.: Smartphone applications targeting precision agriculture practices—a systematic review. Agronomy **10**, 855 (2020). https://doi.org/10.3390/agronomy10060855
7. Buinickaitė, A.: Are large investments necessary to obtain the benefits of precision Farmin. Available online: https://blog.farmis.lt/how-field-navigator-can-help-a-farmer-94aaadf11ae6. Accessed on 25 May 2020
8. Lantzos, T., Koykoyris, G., Salampasis, M.: FarmManager: an Android application for the management of small farms. In: Proceedings of the 6th International Conference on Information and Communication Technologies in Agriculture, Food and Environment (HAICTA 2013), Corfu Island, Greece, 19–22 Sept 2013, pp. 587–592

9. Agroop: Agroop Cooperation—Crop Monitoring. Available online: https://www.agroop.net/en/whatwedo#cooperation. Accessed on 13 Nov 2019
10. AgriApp Technologies: AgriApp—Connecting Farmers. Available online: http://agriapp.co.in/. Accessed on 24 Oct 2019
11. Zargar, H.: AgriApp: An App for farmers to help them improve crop output. Available online: https://www.livemint.com/Technology/btn0QkaCI3rBtdotyiQdfL/AgriApp-An-app-for-farmers-to-help-them-improve-crop-output.html. Accessed on 25 May 2020

Energy Consumption-Based Profiling of Android Apps

Jagannath Singh and Arpan Maity

1 Introduction

As we all know with the growing speed of technological advancement, smart phones have become the essential components of our daily performance. As we look for convenience, we also respect the devices, which can combine multiple features and which give us more mobility and entertainment. As the whole world is going into the new [1] phase of technological performance, our needs become more sophisticated. On the one hand, we need speed, quality, and effectiveness; on the other hand, these features should be combined in a solution small enough to carry it in the pocket. Today, smart phones are the devices which provide all the facilities what a user needs in his daily life, such as e-mail, notebook, Bluetooth, gaming panel, high-resolution camera applications, Microsoft office suite, television and many other computerized applications that a human being can just think of but with a limited battery (power) capacity, and since they are mobile in nature, so cannot be charged regularly [2, 3]. So, we need to optimize the power consumption.

About 80% of the mobile industry uses android as their operating system [1], so there is a flood of android applications in the market. This leads to many applications with the same functionality. Hence, in this competitive market, it is very much important to select the application which is optimal in every way including the power consumed by the application. Through this project, we are trying to get the power consumption of any android application without installing it. The battery drain in smartphones are not only due to faulty batteries or any other hardware issues, but also due to extensive usage, phone's battery seems to be the reason for the battery drain and resolve the issue to get a better user experience.

J. Singh (✉) · A. Maity
KIIT Deemed to be University, Bhubaneswar, India
e-mail: jagannath.singhfcs@kiit.ac.in

J. Singh et al. (eds.), *Mobile Application Development: Practice and Experience*,
Studies in Systems, Decision and Control 452,
https://doi.org/10.1007/978-981-19-6893-8_3

By this project, we are revolutionizing the way android applications are compared in terms of power consumption. Our project will help the user and application designer to test the applications on the basis of power consumed.

2 Basic Concepts

In this section, some basic concepts are presented, which are very relevant and important for understanding our proposed technique.

2.1 Eclipse

Eclipse is an integrated development environment (IDE) used in computer programming and is the most widely used Java IDE. It contains a base workspace and an extensible plug-in system for customizing the environment. Eclipse is written mostly in Java, and its primary use is for developing Java applications, but it may also be used to develop [3] applications in other programming languages via plug-ins, including Ada, ABAP, C, C++, C, COBOL, D, Fortran, Haskell, JavaScript, Julia, Lasso, Lua, NATURAL, Perl, PHP, Prolog, Python, R, Ruby (including Ruby on Rails framework), Rust, Scala, Clojure, Groovy, Scheme, and Erlang.

It can also be used to develop documents with LaTeX (via a TeXlipse plug-in) and packages for the software Mathematica. Development environments include the Eclipse Java development tools (JDT) for Java and Scala, Eclipse CDT for C/C++ and Eclipse PDT for PHP, among others. In Fig. 1, picture of Eclipse dashboard is shown.

The initial codebase originated from IBM Visual Age. The Eclipse software development kit (SDK), which includes the Java development tools, is meant for Java developers. Users can extend its abilities by installing plug-ins written for the Eclipse platform, such as development toolkits for other programming languages, and can write and contribute their own plug-in modules. Since the introduction of the OSGi implementation (Equinox) in version 3 of Eclipse, plug-ins can be plugged-stopped dynamically and are termed (OSGI) bundles. Eclipse software development kit (SDK) is free and open-source software, released under the terms of the Eclipse Public License, although it is incompatible with the GNU General Public License. It was one of the first IDEs to run under GNU Classpath, and it runs without problems under Iced Tea.

2.2 DEX to JAR Convertor

Dex2jar is a lightweight API designed [4] to read the Dalvik executable (.dex/.odex) format. The following steps are used to de-compile the apk!

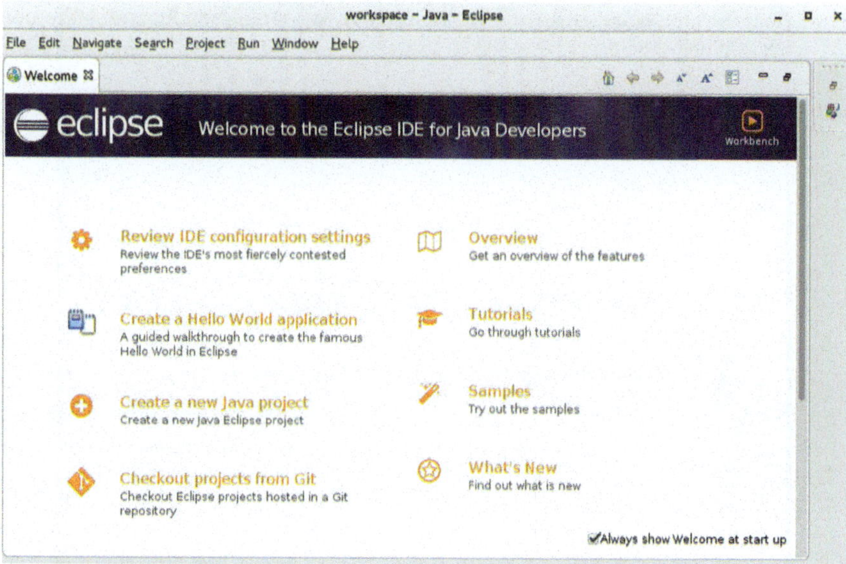

Fig. 1 Eclipse dash board

Step 1: Obtain .apk file. You first need to obtain the .apk file of the application that you wish to decompile.
Step 2: Convert the .apk file into .zip file and extract classes.dex file from it.
Step 3: Using dex to jar convertor convert classes.dex file to .jar file.

2.3 *Bytecode Viewer*

Bytecode Viewer (shown in Fig. 2) is an advanced lightweight Java Bytecode Viewer, GUI Java Decompiler, GUI Bytecode Editor, GUI Smali, GUI Baksmali, GUI APK Editor, GUI Dex Editor, GUI APK Decompiler, GUI DEX Decompiler, GUI Procyon Java Decompiler, GUI Krakatau, GUI CFR Java Decompiler, GUI FernFlower Java Decompiler, GUI DEX2Jar, GUI Jar2DEX, GUI Jar-Jar, Hex Viewer, Code Searcher, Debugger and more.

It is written completely in Java, and it is open sourced. It is currently being maintained and developed by Konloch.

There is also a plug-in system that will allow you to interact with the loaded classfiles; for example, you can write a String deobfuscator, a malicious code searcher, or something else you can think of.

You can either use one of the pre-written plug-ins, or write your own. It supports groovy scripting. Once a plug-in is activated, it will execute the plug-in with a

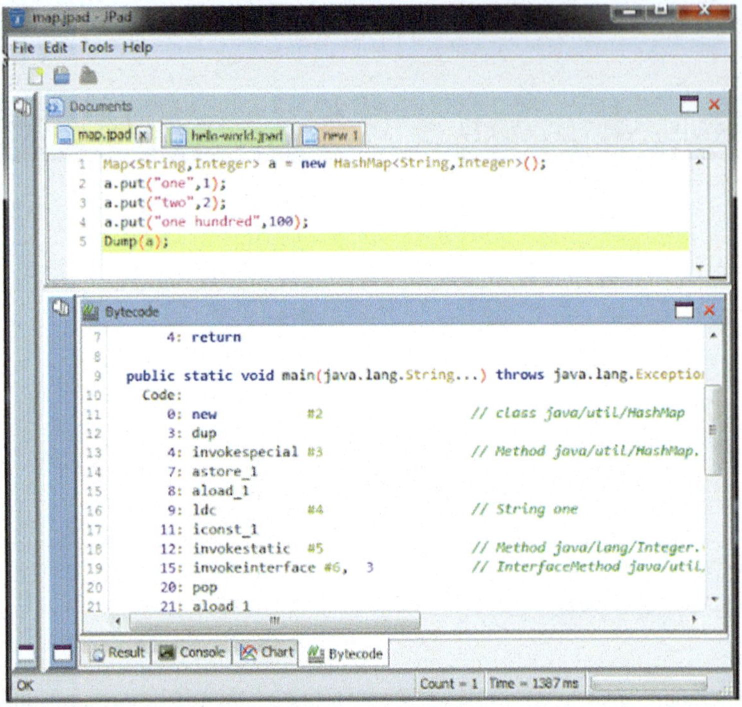

Fig. 2 Bytecode viewer window

ClassNode ArrayList of every single class loaded in BCV, this allows the user to handle it completely using ASM.

3 Related Work

In this section, we review some of the work related to the energy consumption of android apps. In our literature survey, we found that many work has been done in the field of energy consumption of android apps. We have presented few of the closely related works along with their limitations.

In the work of Pinto et al. [6], it is written that out of all the energy requirements of a portable device, the software used most of it. So software optimization is very much important for energy saving. In their paper, power consumption is broadly classified into two groups, processor power and memory power. Energy consumed by the processor mainly depends upon the clock frequency, average supply voltage, average capacitance and node transition activity factor. The flowchart of the algorithm used here is shown in Fig. 3.

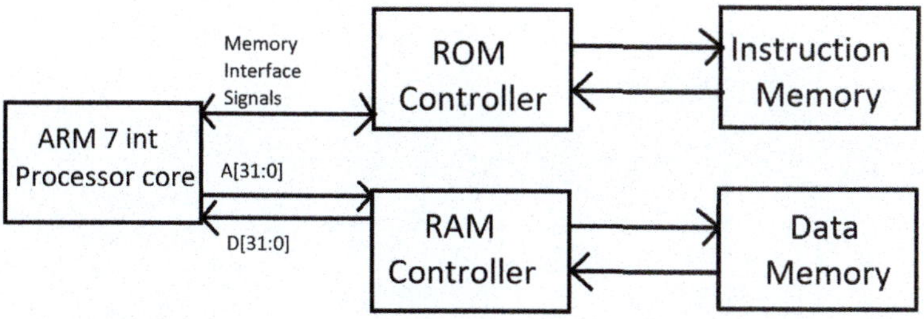

Fig. 3 Flowchart of the algorithm

To find out the energy complexity, the running time of an algorithm was needed and for that assume a generic one-processor, random access machine (RAM) was assumed. Advantage of that kind of model was that there was always a single thread execution, no concurrent execution [6, 7]. So, no context switching took place, and as a result, the energy consumed only to process the algorithm was easy to find.

In the paper titled Estimating Mobile Application Energy Consumption using Program Analysis, written by Hao et al. [4], a very simple method was written, where the energy consumption of the android applications was optimized to a great extent. The method they proposed was both very lightweight in terms of developer requirement, yet was very powerful, so that a visible decrement was observed in terms of energy consumption. It achieves this using a novel combination of program analysis [5] and per-instruction energy modeling. To make the energy consumption optimized, they followed certain steps to reach their goal. At first, they generated their workload. After that, they estimated their energy consumption with the help of eLens, whose structure is shown in Fig. 4. After that they used several energy annotations and then formed the software environment energy profile. This was the overall procedure they used.

After that, we went through the paper of Pinto et al. [6] named mining questions about software energy consumption. In their paper, they did an extensive empirical study on understanding [7] the views of application programmers on software energy consumption problems. They chose their entire problem statement from StackOverflow. Their they studied and analyzed a sample containing 300 questions and 550 answers from more than 800 users. Their main findings are:

I. Energy- and power-related questions have distinct characteristics with respect to the average StackOverflow questions. On an average, they have 2.6 times more answers and, that are marked as favorites 3.89 times more often, have 68% more views, 10% more "up-votes."
II. They identified in total five different themes regarding energy consumption questions.
III. Questions related to energy consumption have a near-linear growth in the last 5 years.
IV. They identified seven major causes for energy consumption and eight solutions to reduce them efficiently.

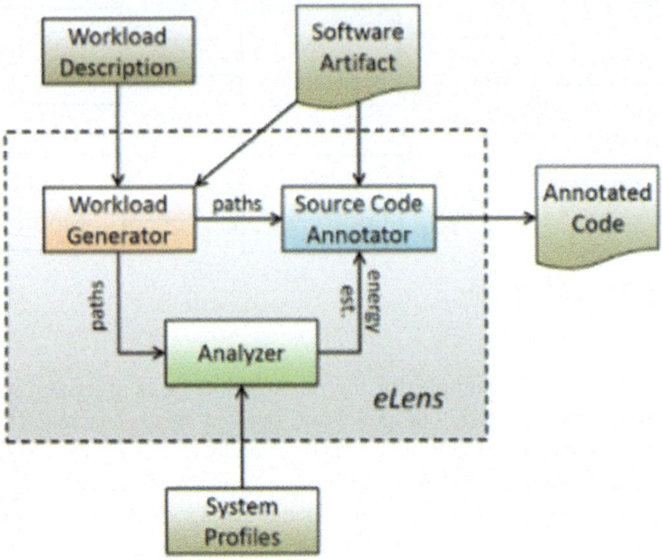

Fig. 4 Structure of eLens

4 Proposed Work

The main key here is an android application. First, we took an android application and changed its extension into zip. Android application is actually a zip file of the source files which an android operating system can read. After converting it into zip, we can now access it with windows/linux and can extract its data [8]. So, after extracting the data, our objective is to find classes.dex file from it. dex file is actually executable form which contains all the classes required for the android application to run on android operating system. Actually, the jar file is converted into dex file during the formation of android application. The detailed algorithm is shown with the help of flowchart in Fig. 5.

After that the dex file is converted into jar file with the help of dex2jar software which is a reverse engineering software from that we got a jar file. With the help of bytecode viewer, we extracted all the .class files which are used in android application making. All the class files are sent to Graphviz (which is software to produce graphs with the help of bytecodes) through java application called dependency analysis [10]. The resultant is a graph for each .class file.

By analyzing the graphs, we can calculate the complexity of the codes by visualising number of cycles and loops. After that, an equation was derived using the complexity[1] as a parameter to get the power consumption by the respective android application.

[1] https://blog.codacy.com/an-in-depth-explanation-of-code-complexity/.

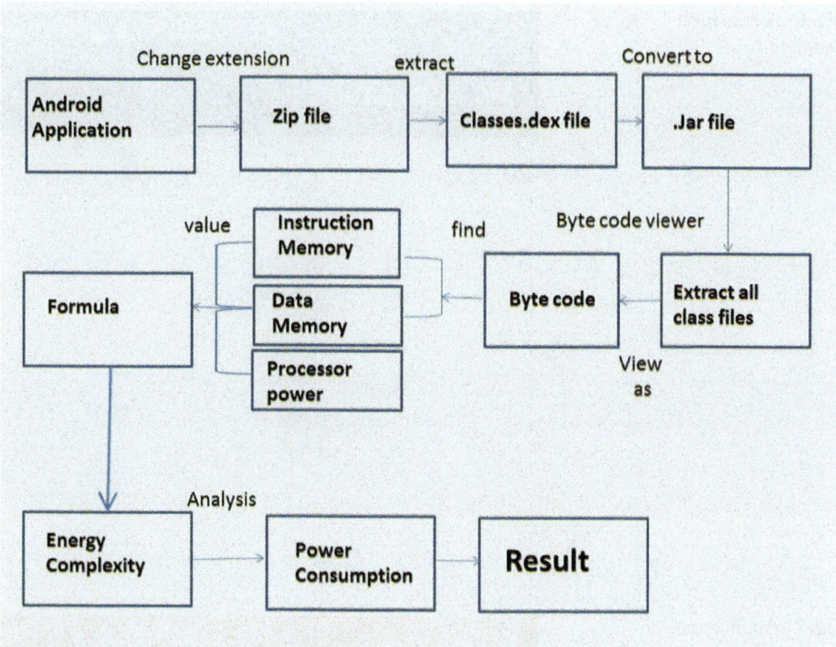

Fig. 5 Workflow of the model used

5 Experimental Setup and Methodology

We are using two demo apps; using their source codes, we will calculate the energy consumption of the apps through a formula and find which one is more efficient.

5.1 Addition

The first application is for addition of two numbers. We can enter two numbers and add them together, and the answer will display on the same screen. The screenshots of the application are shown in Figs. 6, 7 and 8.

5.2 Intent

The second application is simple app to display your name. It will display on the next page.

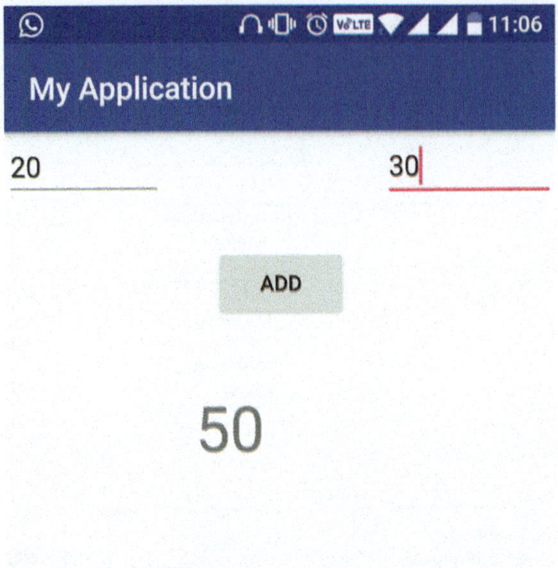

Fig. 6 Application Example 1

Fig. 7 Application Example 2

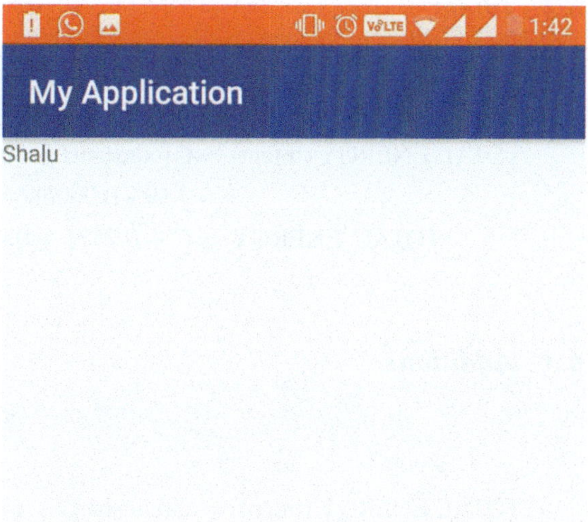

Fig. 8 Application Example 3

6 Methodology

6.1 Formula Derivation

We have used two different android applications and have tried taking out the energy consumption of the respective applications by following the implementation procedure and using the formula derived from the IEEE paper [2, 3]. The formula for that is:

$$\text{Accuracy} = E(\text{processor}) + E(\text{instruction\_memory}) + E(\text{data\_memory})$$

- Processor power is the system power, so we are keeping it as constant.
- Instruction memory is load, store, push, pop, add, sub, goto, return, cmpl, etc.
- Data memory is load, store, push, pop, add, sub, etc.
- And, according to IEEE paper, the energy consumption value is given below:

1. instruction_memory= 0.005142 mJ and
2. data_memory= 0.000686 mJ.

Now for applications—We have taken two android apps.

6.2 *Intent*

$$\begin{aligned}\text{TOTAL ENERGY (intent)} &= C(\text{constant}) + 50 * 0.005142(\text{data\_mem})\\ &\quad +59 * 0.000686(\text{instr\_mem})\\ \text{TOTAL\_ENERGY} &= C + 0.2571 + 0.040474\,\text{mJ}.\end{aligned}$$

6.3 *Additions*

$$\begin{aligned}\text{TOTAL ENERGY (intent)} &= C(\text{constant}) + 44 * 0.005142(\text{data\_mem})\\ &\quad +52 * 0.000686(\text{instr\_mem})\\ \text{TOTAL\_ENERGY} &= C + 0.2571 + 0.040474\,\text{mJ}.\end{aligned}$$

7 Tools Used

7.1 *Eclipse*

See Fig. 9.

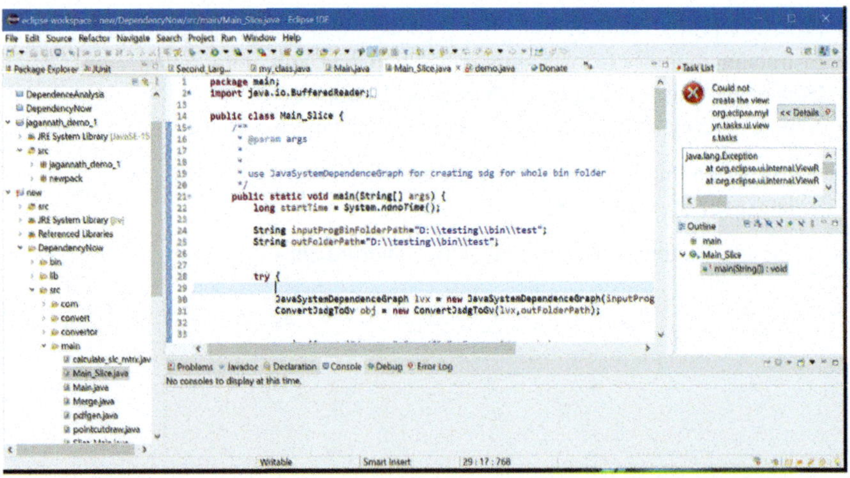

Fig. 9 Homepage of the code in Eclipse

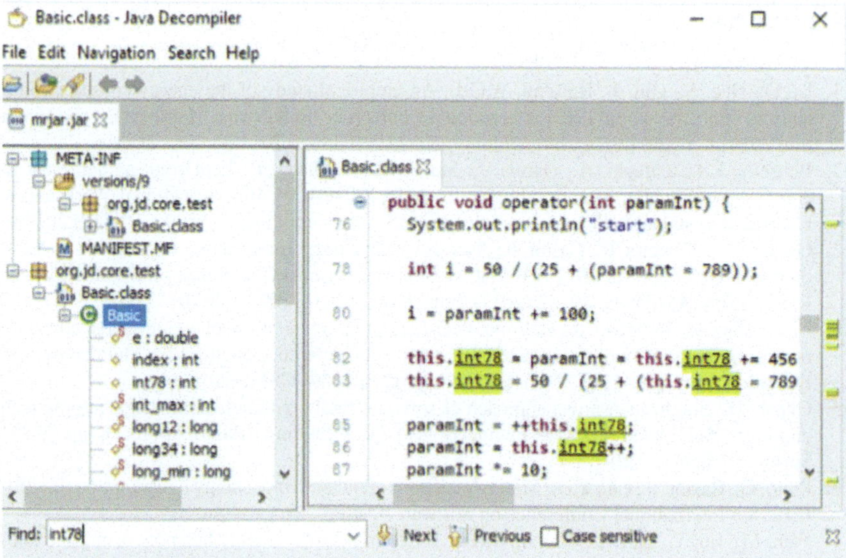

Fig. 10 Bytecode viewer

7.2 *BYTECODE Viewer*

See Fig. 10.

8 Conclusion and Limitations

From the above project, it is clear that we can analyse the power consumption of any android application before installing it. Hence, it is will be very easy for a user to decide between two or more android application having similar function [9], which one will be more power efficient and better to use for the in the long run.

Since android operating system is becoming famous because of its open source api, there are more and more android applications which have the same functions [10]. So, it will be a deciding factor to choose which application is more power efficient. Nowadays, the major concern of any system is battery utilization which can be optimized.

Its major limitation is that it works well only for small-sized applications and programs. Though it is not the fact that it will not work or fail for medium or large sized applications, not much testing is done with those. Our future work will be to test for medium to large apps.

References

1. Li, D., Hao, S., Gui, J., Halfond, W.G.J.: An empirical study of the energy consumption of android applications. In: IEEE International Conference on Software Maintenance and Evolution **2014**, 121–130 (2014). https://doi.org/10.1109/ICSME.2014.34
2. Behrouz, R.J., Sadeghi, A., Garcia, J., Malek, S., Ammann, P.: EcoDroid: an approach for energy-based ranking of android apps. In: 2015 IEEE/ACM 4th International Workshop on Green and Sustainable Software, pp. 8–14 (2015). https://doi.org/10.1109/GREENS.2015.9
3. Oliveira, W., Oliveira, R., Castor, F.: A study on the energy consumption of android app development approaches. In: 2017 IEEE/ACM 14th International Conference on Mining Software Repositories (MSR), pp. 42–52 (2017). https://doi.org/10.1109/MSR.2017.66
4. Li, D., Hao, S., Govindan, R., Halfond, W.G.J.: Estimating mobile application energy consumption using program analysis. In: IEEE International Conference on Software Maintenance and Evolution **2014**, 121–130 (2014). https://doi.org/10.1109/ICSME.2014.34
5. Ghosh, D., Singh, J.: A novel approach of software fault prediction using deep learning technique. In: Automated Software Engineering: A Deep Learning-Based Approach, pp. 73–91. Springer, Cham (2020)
6. Pinto, G., Castor, F., Liu, Y.D.: Mining questions about software energy consumption. In: 2015 IEEE 8th International Conference on Software Testing
7. Wan, M., Jin, Y., Li, D., Halfond, W.G.J.: Detecting display energy hotspots in android apps. In: 2015 IEEE 8th International Conference on Software Testing, Verification and Validation (ICST), pp. 1–10 (2015). https://doi.org/10.1109/ICST.2015.7102585
8. Cruz, L., Abreu, R., Rouvignac, J.: Leafactor: improving energy efficiency of android apps via automatic refactoring. In: 2017 IEEE/ACM 4th International Conference on Mobile Software Engineering and Systems (MOBILESoft), pp. 205–206 (2017). https://doi.org/10.1109/MOBILESoft.2017.21
9. Banerjee, A., Chong, L.K., Ballabriga, C., Roychoudhury, A.: EnergyPatch: repairing resource leaks to improve energy-efficiency of android apps. IEEE Trans. Softw. Eng. **44**(5), 470–490 (2018). https://doi.org/10.1109/TSE.2017.2689012
10. Singh, J., Khilar, P.M., Mohapatra, D.P.: Dynamic slicing of distributed aspect-oriented programs: a context-sensitive approach. Comput. Stand. Interfaces **52**, 71–84 (2017)

Impact of COVID-19 on IT Business

Vishruti Desai, Unnati Shah, Saurya Mehta, Makhania Monil, and Patel Tirth

1 Introduction

Currently, the globe is afflicted by a pandemic that has killed millions of people. COVID-19 has spread over the globe in a relatively short period of time, affecting practically every business. However, over the previous two decades, India's IT sector has seen a rapid expansion. In truth, this industry has been affected by the fatal virus, and there have been several adjustments in employee work patterns and client interactions. Work from home is the new trend for practically all firms, and while there are various drawbacks, there are numerous advantages to working from your most comfortable location, i.e., home.

The vaccination process has begun, but there has been a significant increase in the number of cases that has been detected positive. Various IT corporations have asked their workers to work from home for the next two or three years in order to avoid huge gatherings, as the majority of their work can be accomplished online. As a result of the epidemic, customers have begun to trust Internet services. Working from home appears to be fine, but evaluating the situation on the larger screen COVID-19 has had a significant impact on IT services, either positively or negatively. Furthermore, IT organizations have seen mass job reductions in order to reduce losses, but many employees have suffered as a result. These circumstances have been witnessed all throughout the world, and they have had an impact on their individual countries' financial systems.

V. Desai (✉) · U. Shah · S. Mehta · M. Monil · P. Tirth
Department of Computer Engineering, C. K. Pithawala College of Engineering and Technology, Surat, Gujarat, India
e-mail: vishruti.desai@ckpcet.ac.in

J. Singh et al. (eds.), *Mobile Application Development: Practice and Experience*,
Studies in Systems, Decision and Control 452,
https://doi.org/10.1007/978-981-19-6893-8_4

1.1 COVID-19 in India

Before we can understand the influence of COVID, we must first understand the existing state of cases in the country. The state of Maharashtra in India has the most cases, followed by Kerala and Karnataka. Fever, cough, shortness of breath and breathing difficulty were among the typical symptoms reported. In order to stop the pandemic, India went into a three-month urgent lockdown. However, this infection has continued to spread at a rapid rate, claiming many lives and destroying numerous institutions' crucial services. Many people have gone through financial hardships in order to safeguard their loved ones from this terrible sickness. Furthermore, depression functioned as a catalyst for the situation to worsen even more, as people had no possible work to do during the initial period of lockdown, which had a significant impact on their mental health. People in India are currently feeling relieved after hearing about the effectiveness of vaccines in the country, which is also selling vaccines to other countries in need. Finally, things are returning to normal, but this time, social separation and sanitization have been deemed the most important condition to follow.

1.2 Indian IT Industry

For a few years, the Indian IT industry has been famous for its outstanding software performance and product quality, as well as its timely delivery. Many multinational firms are engaging with Indian enterprises to improve their standards and working methods using cutting-edge technology. With the passage of time, there has been a noticeable increase in technology, which is a gift from globalization and modernity. This improvement has not improved the core IT industry but also has enhanced other industries in which there is a need for some software. Every now and then, people are motivated to start up their own company with some amazing and brainstorming ideas. Entrepreneurship has also been refined in recent years where they offer their employees the perfect stage to show off their real potential which has itself improved the working of IT industries. Talking about interactions with different clients, there are currently many options open to the software industry where they can easily connect with the client and can also assure about the genuinity of the product.

Amidst trying to tackle the difficult situations, India had to struggle with several things including unemployment to financial management. Most companies had to cut off the strength of the number of employees they had which attracted many other dilemmas. The company's talent acquisition function has decided to conduct all interviews through various virtual conference meeting sites like GoogleMeet, Cisco WebEx, Skype, etc. [1].

The COVID-19 impacts on the technology sector, viz. affecting raw materials supply, disrupting the electronics value chain and causing an inflationary risk on products. Favorably, the disruption has accelerated remote working and resulted in a rapid focus on evaluating and de-risking the entire value chain. The majority of businesses do not have a technology stack in place to ensure a viable business continuity plan (BCP). IT departments will play a larger role in future BCPs as a result of improved remote work scenarios and will require assistance from IT service providers in procuring devices, setting up a resilient, flexible and secure network, disaster recovery systems, IT security and other areas [2]. The IT industry is currently experiencing significant weaknesses as a result of the current economic downturn, as many companies are being forced to ask their employees to work from home (remotely) due to public health concerns. Many companies with worldwide dealers are losing a lot of money as a result of this. Apple Inc., for example, is expected to see a 10% drop in its stock due to a scarcity of iPhones on the market. The parts that are required to build the iPhones are supposed to come from China, and it is facing a major lockdown.

Many tech conferences have been canceled as a result of the spread of this fatal illness, which might have been a terrific opportunity for many organizations to extend their horizons through collaboration. A few meetings were moved to teleconferences, but this will not have the same reach, and conference attendees will not be able to network as much as they would if they were at the actual conference. A loss of US$ 1 billion is estimated as a result of the cancelation of these big IT conferences [3].

1.3 Current Challenges for IT Industries [4]

There are surfeit amount of challenges faced by IT industries, and the following are the problems:

1. Organizations have to purchase the needy instruments to work (e.g., laptops). Some required software may not be available or compatible for some of the employees working in that company. Due to this, a lot of operational costs has been increased.
2. Firms also have to provide the facility of Internet used up by employees as everyone might now have enough or smooth Internet facility to carry out their work virtually from home.
3. Those organizations who have taken up space on rent have to pay for it though the place is not utilized for which it was intended to.
4. Employees may have to spend on setting up infrastructure at home so that they can work comfortably, which may add further cost to the employees considering their efficiency remains intact.
5. Internet bandwidth may not be up to mark, which may affect the quality of the delivery.

6. Because the number of employees coming to the office is minimal, the transport requirement is also less. Due to this, the survival of the transport contractor and associated employees is in question because their source income has not been considered in the list of priority. For the IT organization, it will invite risk to retain the vendors and settle down their bills of such contractors which may also lead to some argument.
7. Employees' efficiency may not be up to 100% due to various reasons at home. So, sometimes organizations may have to deal with a degraded performance by adding up some new training seminars which may help employees to gain knowledge about how to cope up with the challenges.
8. Collaborative efforts of teams are lacking currently as coordination between the group members is not constant and effective as compared to the physical workplace which has definitely declined the profits of certain organizations.
9. Critical deadlines might have to be extended further which cause further losses.

2 Data Collection

We have been collecting information from several IT companies over the past few months. The goal of this survey is to determine the extent to which the IT sectors in India have been impacted. Before beginning the survey, we evaluated a number of aspects that could aid us in performing such analyses. We had created a questionnaire with the appropriate set of questions in order to conduct this poll. Google Forms proved to be a really useful tool for conducting this poll, as we were able to send the forms to a variety of IT firms. Here are some of the most critical considerations we made while compiling these questions.

1. Name of the person filling the Google Form.
2. City in which the company is located and the number of full-time employees the company holds.
3. Various issues that employees are facing by working from home.
4. What level of impact do the business has affected to due COVID-19.
5. Modes of interaction with clients (e.g., social media apps: WhatsApp or virtual conference meetings like GoogleMeet or maybe meeting clients in a physical environment.)
6. What type of support does the company want?
7. Quick responsiveness of managers during crisis periods.
8. How much accuracy do employees achieve by working from home?
9. Did the company gain profit or suffer a loss compared to last year?
10. Overall performance of employees has increased or decreased?

We searched for numerous datasets that might be accessible providing the necessary findings in order to find the influence of COVID-19 on IT business. Unfortunately, there was no dataset that met our needs that was published online. As a result, taking the survey was the best option for meeting our requirements.

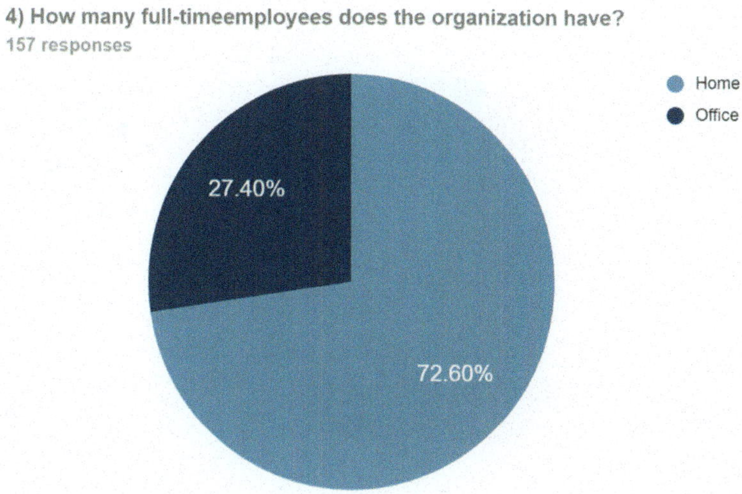

Fig. 1 Working place

3 Result of the Survey

On an active basis, we had been collecting the data from the employees up till now and we have got many analytical outputs based on the input of the survey. Many interesting facts had been discovered in some textual and graphical form. Various visual data had been generated with the help of a Python Jupyter Notebook. Figure 1 shows the results for the working palace.

Even after the lockdown was lifted, the majority of the company continued to operate from home, as shown in Fig. 1. Approximately, 75% of businesses instructed their employees to work from home, ensuring optimal safety for all employees, while nearly 25% of businesses summoned their staff to the office. Coming up with full-time employees, companies had cut off their strength but mostly the companies had enough strength to cover up the expenses and make a profit. Small companies or start-ups employed about 10–30 people, and at the other end, some big companies employed more than 30 people for their organizations to work. Figure 2 shows the results for the *"How many full-time employees does the organization have?"*.

Concerning issues, employees had played a critical role in demonstrating an impact on their business. People encountered a variety of issues, including a lack of resources, financial difficulties for the company, a lack of cooperation between group members on particular projects and irregular working hours, all of which harmed people's efficiency. According to the results of the poll, most employees had trouble coordinating with other team members because they could not reach each other.

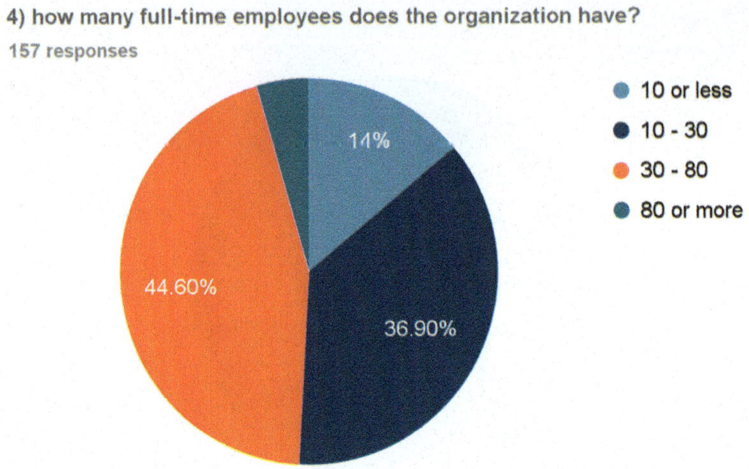

Fig. 2 Number of employees

Figure 3a, b depicts the results of the poll on the disadvantages of working from home.

During such circumstances, businesses maintained contact with their customers through a variety of alternative channels. Virtual conference meetings were the most popular channel for clients to engage, according to the data we collected in the survey. The manager's role is critical in any organization. This fact was, nevertheless, underlined as a result of the survey. The majority of people (about 60%) thought their supervisors were extremely responsive (more than 75%) to any issues the organization faced. Furthermore, the manager's role in generating a set of planning scenarios is very important, and nearly, 60% of businesses were prepared to confront future obstacles and challenges with the most appropriate answer. Figure 4 displays the loss outcome compared to the previous year.

Meeting deadlines is critical for any organization, whether it operates on a large or small basis. If the project is not completed within a certain amount of time, the company may suffer significant losses. Similar instances were noticed in the results of our study, with nearly half of the workers unable to achieve the goal in a set length of time. Figure 5 shows the completed work within the deadline.

When it came to efficiency and performance, most people were uncomfortable working from home because they were not getting the proper environment to work in, and their performance suffered as a result. On the other hand, some people believed that working from home was the ideal place for them to concentrate on their work without feeling pressed, and as a result, their day-to-day performance improved.

(a)

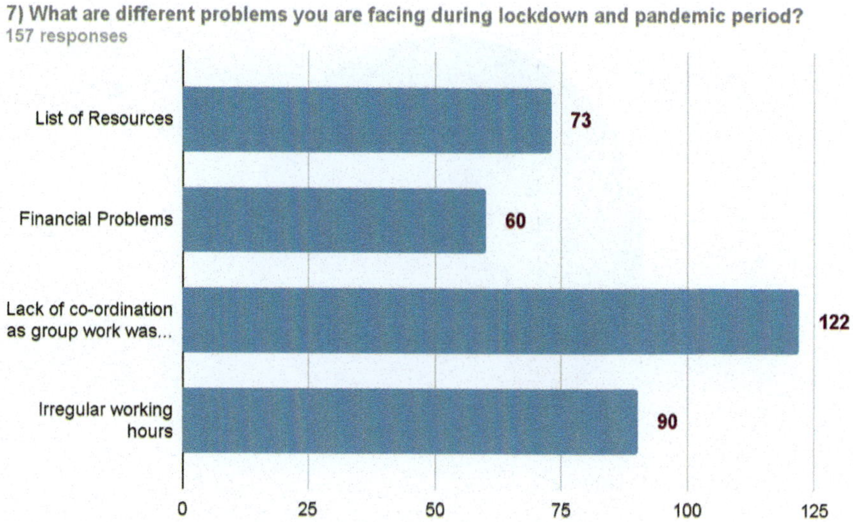

(b)

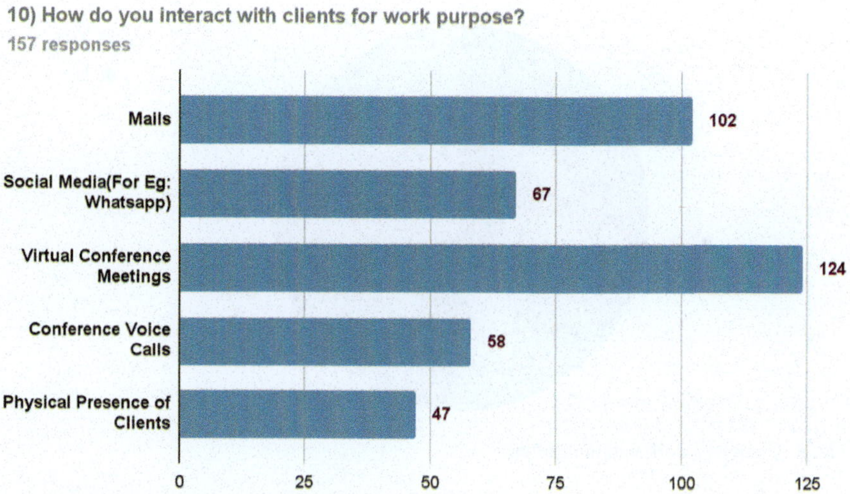

Fig. 3 **a**, **b** Drawbacks to work from home

4 Methodology

The dataset can be divided into many different categories using machine learning techniques; however, in order to work with significant elements, some characteristics were only examined for studying the influence of COVID-19 on the IT business. To

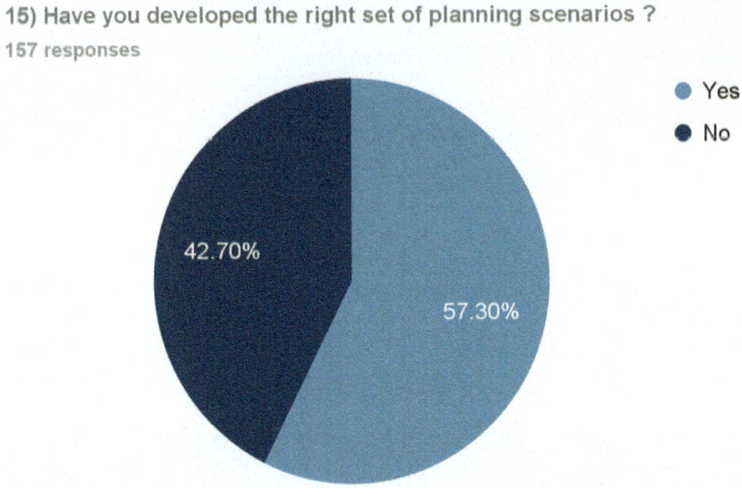

Fig. 4 Loss compared to last year

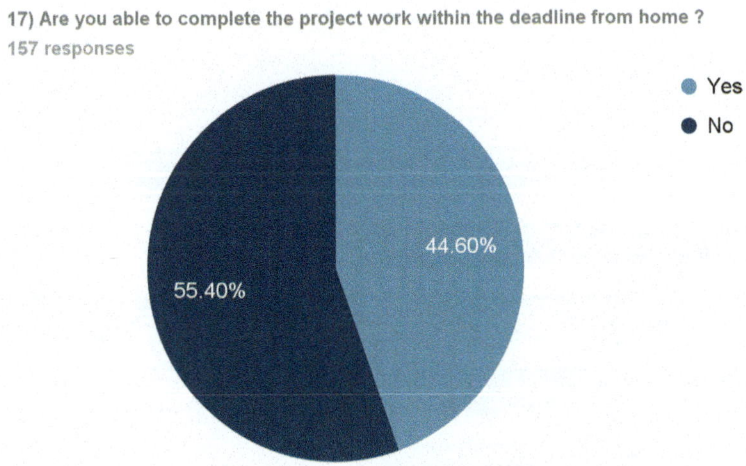

Fig. 5 Complete work within deadline

determine which conditions belonged to which class, the following classification techniques were utilized.

1. Naive Bayes classification.
2. Random forest classification.
3. SVM classification (support vector machine).

The classification on which the classes were divided is as follows:

1. Classification based on profit or loss with respect to loss compared to last year.

2. Classification based on the role of manager which led the business to make profit or loss margin.
3. Classification based on the performance of employee with respect to profit or loss margin. Some basic details of these algorithms are as follows:

1. Naive Bayes Classification:

It is a classification method based on the Bayes theorem and the assumption of predictor independence. A naive Bayes classifier, in simple terms, posits that the existence of one feature in a class is unrelated to the presence of any other feature. The text categorization industry is the primary focus of naive Bayes. It is mostly utilized for clustering and classification purposes, and it is based on the conditional probability of occurrence [5].

Pseudocode of Naive Bayes:

Input: Training dataset T, $F = (f1, f2, f3, \ldots, fn)$ // value of the predictor variable in testing dataset. Output: A class of testing dataset. Steps: (1) Read the training dataset T; (2) Calculate the mean and standard deviation of the predictor variables in each class; (3) Repeat and calculate the probability of fi using the gauss density equation in each class; Until the probability of all predictor variables $(f1, f2, f3, \ldots, fn)$ has been calculated. (4) Calculate the likelihood for each class; (5) Get the greatest likelihood [6].

2. Random Forest Classification:

A random tree is one that is generated at random from a set of possible trees, each with K random features at each node. In this context, "at random" means that each tree in the set has an equal probability of being sampled. Alternatively, trees can be described as having a "uniform" distribution. Random trees may be constructed quickly, and combining huge sets of random trees produces realistic models in most cases. In the field of machine learning, there has been a lot of research on random trees in recent years [7].

3. SVM Classification (Support Vector Machine):

Support vector machine is another popular state-of-the-art machine learning approach (SVM). Support vector machines are supervised learning models with related learning algorithms for classification and regression analysis in machine learning. SVMs may perform nonlinear classification as well as linear classification by implicitly mapping their inputs into high-dimensional feature spaces, which is known as the kernel trick. It essentially defines the boundaries between the classes. The margins are drawn so that the space between the margin and the classes is as little as possible, reducing the classification error [8].

Pseudocode of Support Vector Machine Initialize $Yi = YI$ for $i \in I$ repeat compute svm solution vv, b for data set with imputed labels compute outputs $ii = (vv, xi) + b$ for all xi in positive bags set $yi = \text{sgn}(fi)$ for every i in I, $yi = 1$ for (every positive bag bi) end if (liei($l + yi$)/2 $= = 0$) compute $i* = \arg$ maxi ii set $yi* = 1$ end while (imputed labels have changed) output (vv, b) [6].

5 Result of Classification

Here as mentioned above, we had classified data with certain conditions which can be seen through the following pie charts (Figs. 6 and 7).

Police and medical services are almost identical to IT services. If we do not deliver to our clients on schedule, it will have a significant negative impact. Because of COVID-19, multiple stakeholders will need to work together to handle the conflict situation. The Indian government will encourage state governments to instruct IT companies to allow their staff to work from home (It is in place already). Currently, the majority of organizations are doing so. IT companies follow the Government of India's different advices and assist the country throughout this crisis. IT organizations make the appropriate arrangements, such as purchasing laptops and data cards, allowing employees to use their own systems for delivery, and ensuring network security, among other things. Authorities in charge of electricity supply must maintain

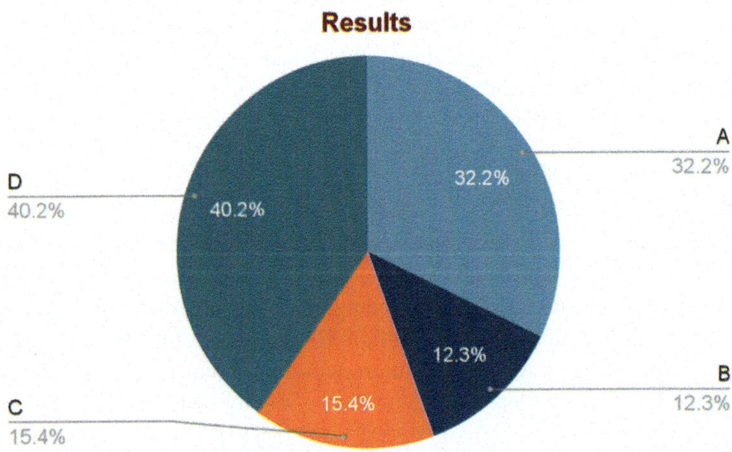


```
df['result'] = 'A'
for 1 in range (len(df):
if df.loc[1,'Loss compared to last year'] == 'Yes' and df.loc[1
    ,'profit /Loss(%)] == '<25%':
df['result'] ='A'
elif df.loc[1,'Loss compared to last year'] == 'Increased' and df
    .loc[1,'profit /Loss(%)] == '>25%':
df['result'] ='B'
elif df.loc[1,'Loss compared to last year'] == 'Decreased' and df
    .loc[1,'profit /Loss(%)] == '>25%':
df['result'] = 'C'
else df.loc['result'] = 'D'
```

Fig. 6 Loss compared to last year

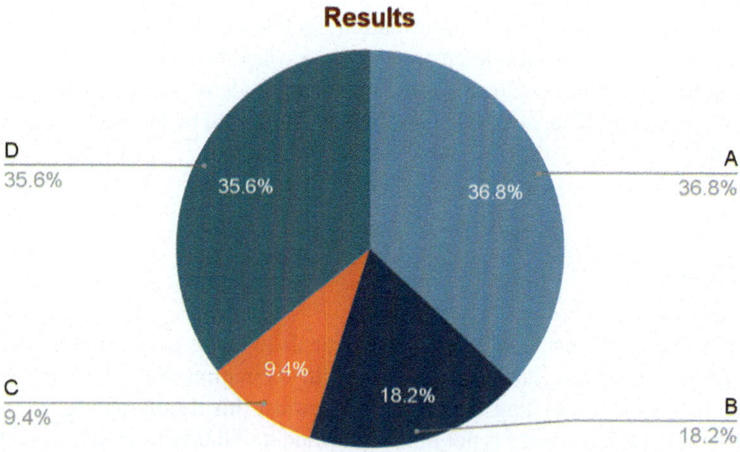


```
df['result']='A'
for 1 in range (len(df):
if df.loc[1,'performance']=='Increased' and df.loc[1,'profit /Loss(%
     )]=='<25%':
df['result']='A'
elif df.loc[1,'performance']=='Increased' and df.loc[1,'profit /Loss
     (%)]=='>25%':
df['result']='B'
elif df.loc[1,'performance']=='Decreased' and df.loc[1,'profit /Loss
     (%)]=='>25%':
df['result']='C'
else df.loc['result']='D'
```

Fig. 7 Performance

an uninterrupted supply so that employees can work without interruption. Telecom providers must guarantee a proper mobile network and adequate bandwidth so that employees may work without interruption. Every IT person is available for the business as a responsible employee. Though it is an extra cost, having a UPS connection and several data connections is required so that consumers are not inconvenienced. Indian IT organizations should collaborate with a variety of stakeholders, including other delivery centers in different countries and delivery centers within India where COVID-19 has a lower influence so that alternative tactics can be devised. Employees must have a good working relationship. During this critical scenario, the relationship between the team manager and the employees is crucial. At this critical juncture, the employer–employee relationship is critical. A backup contact/mobile number/address database should be kept on hand so that the management or employer may reach out to staff in an emergency. Employees, supervisors and executives at all levels of Indian IT organizations should provide open and honest feedback and suggestions. They must create policies and processes for improved implementation using such inputs. The Ministry of Information Technology will collaborate with telecom authorities to limit mobile customers' data usage and only allow them to use it for necessary purposes. Downloading music and videos, as well as watching HD movies, use a lot of data, and hence, these features should be blocked. So, at the very least, the [4].

6 Conclusion

To summarize, this approach enables IT professionals to make swift business decisions based on the present state of many firms in their immediate environment. Vaccination is presently assisting us in the fight against this deadly pandemic, but the impact on various businesses is not immediate, and it will take time to restore the firm to its previous state. This concept will assist those who are considering becoming future entrepreneurs in building their businesses by analyzing their specific location and many factors such as the number of employees and their performance from either home or office. It is clear that COVID-19 has had an impact on all enterprises, either positively or negatively. However, thanks to recent advancements in digitization, Indian IT companies are growing at a rapid pace. While there may be some difficulties in the early stages of setting up, they can eventually stabilize the situation by adhering to particular routines. This is only a notion that has been used in IT disciplines, but it can be applied to a range of various enterprises with the correct set of research in different domains. To start a business, all one needs to do is ask a reasonable set of questions to a businessperson about the difficulty they are facing. We conducted research and identified the present issues that employees are experiencing; if these issues are addressed, the business would undoubtedly prosper in the near future.

References

1. Growth a big challenge for India IT industry in 2020–21 due to coronavirus impact: Infosys CFO. Retrieved from https://www.thehindubusinessline.com/info-tech/growth-a-big-challenge-for-india-itindustry-in-2020-21-due-to-coronavirus-impact-infosys-cfo/article31040354.ece
2. Narayanamurthy, G., Tortorella, G.: Impact of COVID-19 outbreak on employee performance—moderating role of industry 4.0 base technologies. Int. J. Prod. Econ. **234**, 108075 (2021)
3. Shankar, K.: The impact of COVID-19 on the IT services industry-expected transformations. Br. J. Manag. **31**(3), 450 (2020)
4. Ramasamy, D.: The challenges in the Indian IT industry due to COVID-19—an introspection (2020)
5. Rish, I.: An empirical study of the naive Bayes classifier. In: IJCAI 2001 Workshop on Empirical Methods in Artificial Intelligence, vol. 3, no. 22, pp. 41–46 (2001)

6. Mahesh, B.: Machine learning algorithms—a review. Int. J. Sci. Res. (IJSR) **9**, 381–386 (2020)
7. Ali, J., Khan, R., Ahmad, N., Maqsood, I.: Random forests and decision trees. Int. J. Comput. Sci. Issues (IJCSI) **9**(5), 272 (2012)
8. Yue, Y., Finley, T., Radlinski, F., Joachims, T.: A support vector method for optimizing average precision. In: Proceedings of the 30th Annual International ACM SIGIR Conference on Research and Development in Information Retrieval, pp. 271–278 (2007)

ANNDroid: A Framework for Android Malware Detection Using Feature Selection Techniques and Machine Learning Algorithms

Arvind Mahindru

1 Introduction

Android has gained popularity in the year 2011 due to its open-source and number of free apps in its official play store.[1] According to the statistics,[2] more than 2.87 million free apps are present in Google Play Store. Working of android apps depends upon the permissions. At the time of installation, android apps required certain permissions that are required for its proper functioning. On daily basis, cyber-criminals are taking advantage of these permissions and develop malware-infected apps for smartphone users. According to the survey done by Kaspersky Security Network,[3], there are millions of malware-infected apps which are still submitted in Google Play Store and third-party app stores.

According to the report published by Gartner,[4], the growth of smartphone is increased by 11% in the upcoming year. During pandemic, everyone dependent upon apps for their jobs. At the time of installation and run-time, android apps demand certain permissions. Google defined these permissions[5] as "normal" or "dangerous." Normal permissions do not pay any impact on user's privacy. In the reverse, dangerous permissions paid a great effect on user's privacy. The fault lies in the underneath permission model of android apps.

In the literature [12, 14–24], number of authors proposed android malware detection frameworks using supervised and unsupervised machine learning techniques.

[1] https://play.google.com/store.
[2] https://buildfire.com/app-statistics/#.
[3] https://securelist.com/ksb-2020/.
[4] https://indianexpress.com/article/technology/tech-news-technology/smartphone-sales-expected-to-grow-by-11-in-2021-5g-phones-to-play-key-role-7175925/.
[5] https://developer.android.com/guide/topics/permissions/overview.

A. Mahindru (✉)
Department of Computer Science and Applications, D.A.V. University, Jalandhar, India
e-mail: er.arvindmahindru@gmail.com

J. Singh et al. (eds.), *Mobile Application Development: Practice and Experience*, Studies in Systems, Decision and Control 452,
https://doi.org/10.1007/978-981-19-6893-8_5

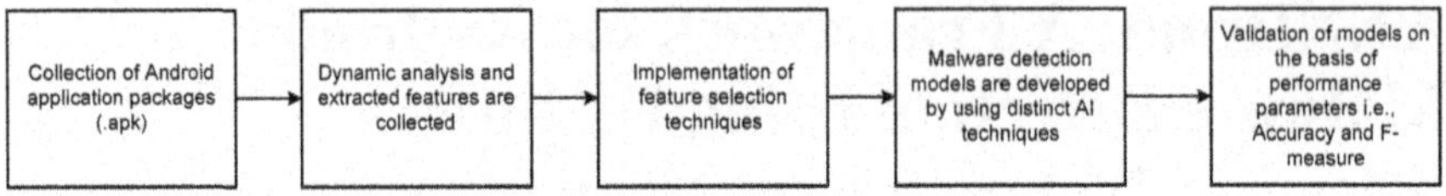

Fig. 1 Phases involved in this research article

The main limitation in their work is that researchers and academicians used limited datasets. In order to achieve better detection rate, in this research article, we proposed a framework that is based on the principle of hybrid artificial intelligence techniques approach of functional link artificial neural network (FLANN) with clonal selection algorithm (CSA), particle swarm optimization (PSO) and genetic algorithm (GA), i.e., FLANN-CSA (FCSA), FLANN-PSO (FPSO and MFPSO) and FLANN-genetic (FGA and AFGA). This study also focuses on the effectiveness of feature selection techniques, i.e., principal component analysis (PCA) and rough set analysis (RSA), which are used to reduce the complexity of the proposed model by minimizing the number of inputs.

The generic steps that are followed in this research paper to identify malware-infected apps are shown in Fig. 1. Initially in the first step, we collect Android Application Packages (.apk) files from different repositories. In the second step, we extracted dynamic features and form the features dataset. Implemented of feature selection techniques is performed in the third step. Further, features are selected by implementing feature selection approaches. In the last step, we validate our developed models by using two performance parameters, i.e., accuracy and F-measure.

The unique and novel contributions of this study are as follows:

- To build efficient and effective malware detection model, in this study more than five millions android apps are utilized.
- Dynamic analysis was performed on collected android apps, and 1844 unique features are extracted.
- In this chapter, five different hybrid functional link artificial neural networks are proposed.

The rest of the chapter is summarized as follows. In Sect. 2, related work is discussed. Collection of .apk file and formulation of feature dataset is discussed in Sect. 3. Implemented feature selection techniques are discussed in Sect. 4. Section 5 discusses the proposed hybrid machine learning algorithms. Experimental setup to proposed the framework is discussed in Sect. 6. Outcome of the experiment is discussed in Sect. 7. At last, chapter is concluded in Sect. 8.

2 Related Work

Hou et al. [7] proposed a malware detection framework named as "Droiddelver" based on Application Programming Interface (API) that is extracted from smali files. Proposed model was build by using 5000 different android apps and a deep belief network as a machine learning technique. Empirical outcome reveals that the proposed model was able to detect 96.66% of malware-infected apps. Hou et al. [6] proposed a malware detection model named as "Deep4MalDroid" developed on the basis of dynamic analysis approach called component traversal which follows code routines of particular android apps. Based on the extracted features, they construct the weighted directed graphs and then applied deep learning as a machine learning algorithm. An experiment was performed by using 3000 android apps and detect 91.4% malware-infected apps.

Mahindru and Singh [25] proposed dynamic analysis-based approach that are build by using 123 features. An experiment was performed by using 11,000 distinct android apps and five different machine learning algorithms. The malware detection model developed by using simple logistic achieved a higher detection rate as compared to others. Hou et al. [8] developed a framework entitled as "HinDroid" based on the relationships between API calls and developed higher-level semantics that require more efforts for attackers. An experiment was performed by using two different datasets; i.e., one contains 1834 distinct android apps, and the second contains 30,000 distinct android apps. Proposed malware detection framework was able to identify 99.01% malware-infected apps. Martín et al. [26] developed a model named as "MOCDroid," that is based on the integrity of genetic algorithm. An experiment was performed by using 17,135 android apps and achieved an accuracy of 94.60%. Tong and Yan [30] proposed a hybrid approach that works on the combination of static and dynamic features. Experiment was carry-out by utilizing 2000 different android apps while considering API calls as a feature. Proposed malware detection model achieved the detection rate of 90.19%.

Karbab et al. [10] developed malware detection model named as "MalDozer" that is based on the principle of deep learning techniques. Developed model uses the behavior of API calls to recognize the behavior of benign and malware apps. The developed framework was tested on 38 K benign apps and 33 K malware-infected apps and achieved an $F1$-score of 96–99%. Cai et al. [4] proposed a dynamic malware detection approach that used calls and inter-component communication as features. An experiment was performed by using 34,343 android apps and the proposed framework achieved an accuracy of 97%. Kim et al. [11] developed a malware detection model on the basis of multimodal deep learning. Features were extracted from the manifest file, dex file and shared libraries for developing the model. The developed model was tested with 41,260 android apps and achieved an accuracy of 98%. Yerima et al. [34] proposed detection model entitled as "DroidFusion," that is based on the principle of feature selection techniques and implement multiple machine learning algorithms. The proposed malware detection model was tested with 55,018 distinct smartphone apps and achieved the detection rate of 97%. Shen et al. [28] developed a malware detection model that works on the principle of information flow analysis. Developed model is based on the structure of information flows to know the pattern behavior and which helps in distinguishing between benign and malware app. An experiment was performed by using 8598 android apps and achieved an accuracy of 82%.

Arora et al. [1] developed malware detection framework work on graphs that construct by utilizing permissions extract from distinct android apps. An experiment was performed by using 5993 android apps and achieved the detection rate of 95.44%. Xiao et al. [32] developed a model by using deep learning principles. The proposed model is built by using system call sequences and long short-term memory as a machine learning technique. An experiment was performed by using 7103 android apps and achieved an accuracy of 96.6%. Mahindru and Sangal [14] developed a malware detection framework entitled as "DeepDroid" by using significant features selected by feature selection approaches and deep learning as a machine learning technique. Experimental outcome reveals that the framework build by using principal component analysis (PCA) as a feature selection technique achieved a higher detection rate as compared to other techniques. Kumar et al. [12] build a detection framework by utilizing three different data sampling approaches, three different feature selection approaches and seven distinct classifier approaches. Outcome reveals that the framework developed by using upscale sampling technique and ELM with polynomial kernel achieved a higher detection rate as compared to others.

Mahindru and Sangal [16] developed a malware detection framework work on the basis of semi-supervised machine learning techniques. The proposed framework is developed by using four different feature subset selection approaches and LLGC as a machine learning algorithm. The empirical result reveals that framework build using rough set analysis as a feature selection approach achieved the detection rate of 97.8%. Mahindru and Sangal [17] developed malware detection model entitled as "GADroid" that is build by using genetic algorithm as a feature selection approach. Further, selected features are used to build the model by using deep learning as machine learning technique. Experiment was performed on 560,142 distinct android apps, and the developed model is able to achieved an accuracy of 98.6%.

Mahindru and Sangal [19] developed the model named as "PARUDroid." Proposed model is able to detect 98.8% malware-infected apps. Table 1 describes the frameworks developed in the literature. Previous malware detection model has been proposed with a limited dataset and conquered a higher accuracy with the limited dataset. On the basis of related work, the following questions have been answered in this research article:

RQ1. To identify which malware detection model is more effective in detecting malware from real-world apps?

This question helps in identifying the malware detection model which is more effective in detecting malware from real-world apps. To answer this question, in this study distinct malware detection models are developed and compared with two different performance parameters, i.e., F-measure and accuracy.

RQ2. Is the proposed malware detection framework able to identify malware from android devices or not?

Table 1 Malware detection frameworks that are availables in the literature

Framework	Machine learning algorithm implemented	Dataset used
Droiddelver [7]	Deep neural network	6000
Deep4MalDroid [6]	Deep neural network	3000
HinDroid [8]	Heterogeneous information network	31,834
MalDozer [10]	Deep neural network	71 K
DeepDroid [14]	Deep neural network	120,000
GADroid [17]	Deep neural network	560,142
PARUDroid [19]	Deep neural network, decision tree Adaboost, Naïve Bayes and random forest	560,142
DLDroid [15]	Deep neural network	11,000
PerbDroid [20]	SVM, Naïve Bayes, random forest, multiple layer perceptron, logistic regression, Bayesian network, Adaboost, decision tree, *K*NN and deep neural network	200,000
MLDroid [18]	SVM, Naïve Bayes, random forest, logistic regression, multiple layer perceptron, *k*-nearest neighbors, Adaboost, self-organizing map, Bayesian network, deep neural network, decision tree, *K*-mean, density-based clustering, filtered clustering, farthest first clustering, MLP + YATSI, J48 + YATSI, SMO + YATSI, best training ensemble approach, majority voting ensemble approach and nonlinear ensemble decision tree forest approach	
SemiDroid [21]	Farthest first clustering, *K*-mean, self-organizing map, filtered clustering, density-based clustering,	550,000
SOMDroid [22]	Self-organizing map	500,000
FSDroid [23]	LSSVM with linear polynomial and radial kernel	200,000
HybriDroid [24]	Best training ensemble majority voting ensemble and nonlinear ensemble decision tree forest	194,659 benign apps and 67,538 malware apps

To examine this question, in this study, proposed framework is compared with existing malware detection models presented in the literature.

RQ3. While selected features using feature selection approaches paid any impact on malware detection models or not?

To answer this question, developed using in this research article model developed using all extracted features compared with the models developed by using feature selection techniques.

3 Collection of *.apk* Files and Formulation of Features Dataset

Collection of five million distinct android apps is performed to use in this research article. Benign .apk files are collected from, i.e., slideme,[6] mumayi,[7] hiapk,[8] appchina,[9] Google's play store,[10] Android,[11] gfan,[12] and pandaapp,[13] and malware-infected apps are collected from Android Malware Genome project [35], 1929, botnet samples were collected from [9] and from AndroMalShare[14] along with their package names. Table 2 represents the distinct categories of android apps with respect to its numbers. Dynamic analysis was performed by using the principle mentioned in [19]. After that, we divided the extracted features into different categories to which they belong. Formulation of feature dataset is mentioned in Table 3.

4 Feature Selection Techniques

Relevant features paid an important role while developing the malware detection models in case of effectiveness and efficiency. In this research article, to select relevant features two different feature selection approaches are considered, i.e., principal component analysis (PCA) and rough set theory.

4.1 Principal Component Analysis (PCA)

To carry-out a data space, low dimension PCA is considered as feature selection. Figure 2 demonstrates the steps that are considered while selecting features using PCA.

---

[6] http://slideme.org/.

[7] http://www.mumayi.com/.

[8] http://apk.hiapk.com/.

[9] http://www.appchina.com/.

[10] https://play.google.com/store?hl=en.

[11] http://android.d.cn/.

[12] http://apk.gfan.com/.

[13] http://download.pandaapp.com/?app=soft&controller=android#.V-p3f4h97IU.

[14] http://202.117.54.231:8080/.

Table 2 Collected android application packages (*.apk*)

ID	Category	Normal	Trojan	Backdoor	Worms	Botnet	Spyware
DS1	Arcade and action (AA)	15,291	13,300	5000	1004	3300	5000
DS2	Books and reference (BR)	16,235	8000	6500	4060	5650	1600
DS3	Brain and puzzle (BP)	13,928	10,820	204	2008	5010	5010
DS4	Business (BU)	18,208	1420	1150	3250	1842	1602
DS5	Cards and casino (CC)	12,786	7610	6520	8002	5010	4220
DS6	Casual (CA)	16,000	8270	8080	6052	7840	5840
DS7	Comics (CO)	20,967	6050	9950	9900	9900	6100
DS8	Communication (COM)	76,309	8503	5007	8904	8791	8020
DS9	Education (ED)	38,764	8610	8121	8980	8219	8021
DS10	Entertainment (EN)	23,988	8100	8100	7000	1870	5397
DS11	Finance (FI)	23,099	8990	7609	9199	6985	9012
DS12	Health and fitness (HF)	18,661	9181	6852	4825	1840	1940
DS13	Libraries and demo (LD)	13,755	1479	1989	1300	6291	6900
DS14	Lifestyle (LS)	19,650	1855	9805	1808	1093	5082
DS15	Media and video (MV)	18,119	7807	7023	8662	2450	6971
DS16	Medical (ME)	36,000	1128	1983	2344	2884	4805
Ds17	Music and audio (MA)	27,057	6935	5900	6125	1165	2665
DS18	News and magazines (NM)	28,164	4500	3100	2100	1100	1032
DS19	Personalization (PE)	14,334	1580	1042	2590	4280	2170
DS20	Photography (PH)	19,033	3109	4190	2850	9161	5200
DS21	Productivity (PR)	19,750	3600	8903	4350	3290	2972
DS22	Racing (RA)	23,766	1458	3109	4219	8190	2189
DS23	Shopping (SH)	14,673	3120	1950	3120	3150	1959
DS24	Social (SO)	36,159	3190	4550	1210	5159	7159
DS25	Sports (SP)	32,669	6100	7249	9180	4490	8022
DS26	Sports games (SG)	31,889	9200	8045	8125	8250	9198
DS27	Tools (TO)	25,646	9720	8844	7259	9205	4763
DS28	Transportation (TR)	23,796	3102	4200	9100	8002	4120
DS29	Travel and local (TL)	38,180	9508	8220	7050	1248	8100
DS30	Weather (WR)	20,841	7190	2323	9790	3950	2925

4.2 Rough Set Theory

Rough set theory used to eliminate irrelevant features by using approximation, reduced attributes and information method. Steps that are followed in rough set theory are shown in Fig. 3.

Table 3 Formulation of feature datasets

Set number	Description	Set number	Description
FS1	Contain info. Associated to rating and downloads	FS2	Associated to SMS_MMS
FS3	Associated to IMAGE	FS4	Associated to HARDWARE_CONTROLS
FS5	Associated to READ	FS6	Associated to BROWSER_INFORMATION
FS7	Associated to WIDGET	FS8	Associated to SYSTEM_SETTINGS
FS9	Associated to CONTACT_INFORMATION	FS10	Associated to FILE_INFORMATION
FS11	Associated to default group	FS12	Associated to LOCATION_INFORMATION
FS13	Associated to BUNDLE	FS14	Associated to CALEN-DAR_INFORMATION
FS15	Associated to SYNCHRONIZATION _DATA	FS16	Associated to DATABASE_INFORMATION
FS17	Associated to READ_AND_WRITE	FS18	Associated to UNIQUE_IDENTIFIER
FS19	Associated to LOG_FILE	FS20	Associated to ACCOUNT_SETTINGS
FS21	Associated to PHONE_CALLS	FS22	Associated to ACCESS_ACTION R
FS23	Associated to SER-VICES_THAT_COST_YOU_MONEY	FS24	Associated to SYSTEM_TOOLS
FS25	Associated to YOUR_ACCOUNTS	FS26	Associated to NETWORK_INFORMATION and BLUE-TOOTH_INFORMATION
FS27	Associated to AUDIO and VIDEO	FS28	Associated to PHONE_STATE and PHONE_CONNECTION
FS29	Contain info. Associated to API calls	FS30	Associated to STORAGE_FILE

5 Proposed Hybrid Machine Learning Techniques

In this section, we discuss various machine learning algorithms that are developed by using genetic algorithm, clonal selection and particle swarm optimization for detection malware from android apps.

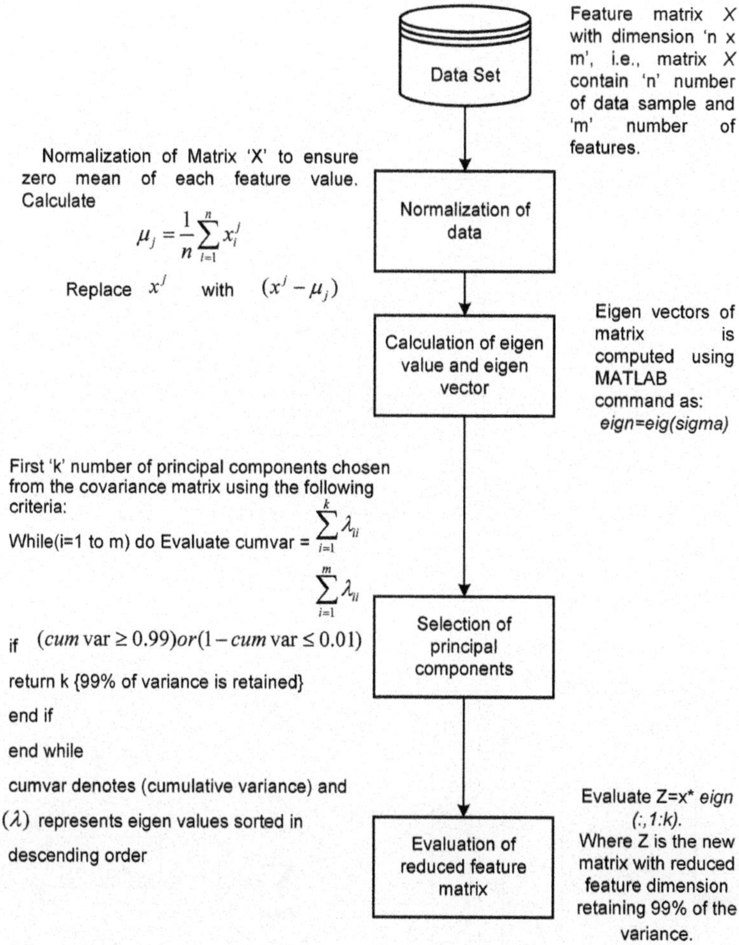

Fig. 2 Framework of PCA calculation

5.1 Functional Link Artificial Neural Network (FLANN)

In this research article, FLANN is implemented to detect malware from android apps. FLANN is worked on the architecture of single layered of artificial neural network (ANN), that is responsible to perform complex decision. The computational cost of ANN is very high, but in the case of FLANN it is very less due to not present of hidden layers. Figure 4 demonstrate the basis architecture of FLANN.

Output is computed by using following equations:

$$\hat{z} = \sum_{i=1}^{n} W_i a_i \tag{1}$$

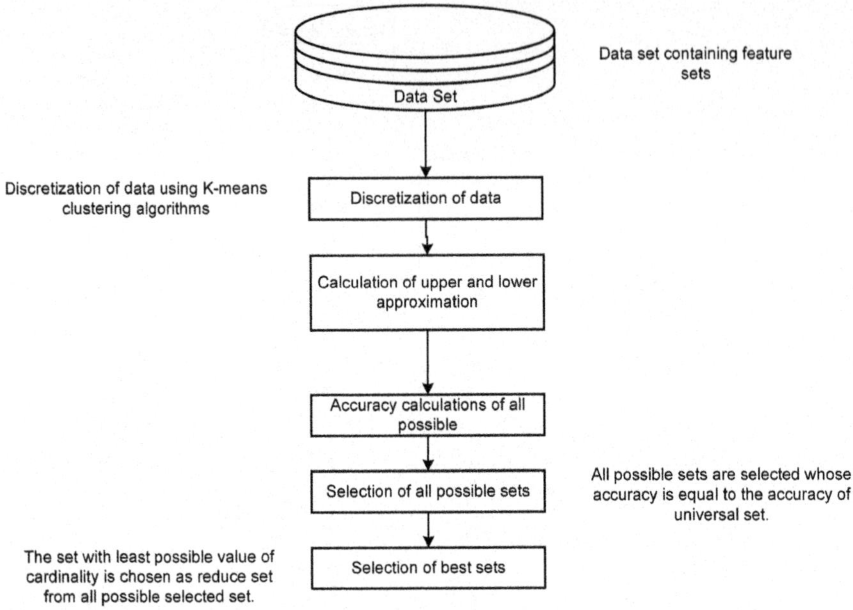

Fig. 3 Rough set theory framework

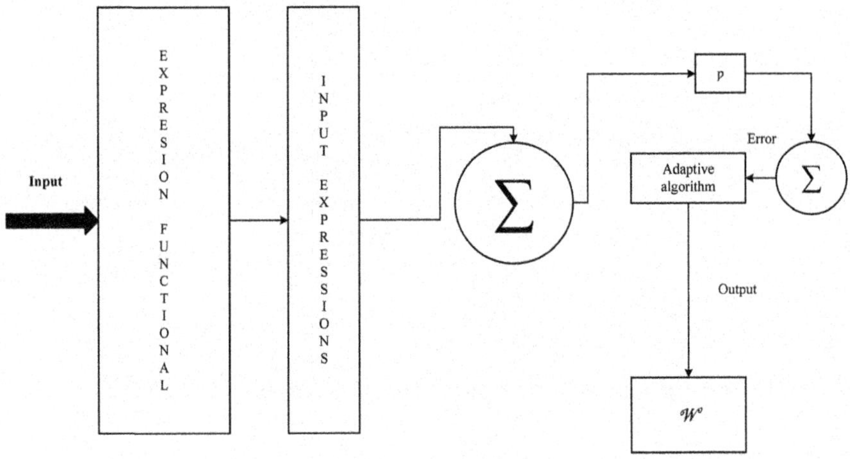

Fig. 4 Architecture of FLANN

where z and $\hat{z}$ are the estimated and actual values, a_i is the function block and W is the weight vector that is defined by using

$$A = [1, a_1, \sin \pi a_1, \cos \pi a_1, a_2, \sin \pi a_2, \cos \pi a_2, \ldots] \tag{2}$$

The revised weight is updated as:

$$W_i(k+1) = W_i(k) + \alpha e_i(k) a_i(k) \tag{3}$$

where e_i is the error value and α is the learning rate that is determined as:

$$e_i = z_i - \hat{z}_i \tag{4}$$

5.2 *FLANN-Genetic (FGA) Technique*

This technique is very effective at the time of learning, and it is utilized mostly there for upgrading the weight. A function link neural network with a form of '$a - x$' is deemed as estimation; i.e., the network contains l number of input neurons and x number of output neurons.

Weights are calculated using the following equation:

$$W_a = \begin{cases} -\frac{y_{ad+2}*10^{d-2}+y_{ad+3}*10^{d-3}+\cdots+y_{(a+1)d}}{10^{d-2}} & \text{if } 0 \le y_{ad+1} < 5 \\ \frac{y_{ad+2}*10^{d-2}+y_{ad+3}*10^{d-3}+\cdots+y_{(a+1)d}}{10^{d-2}} & \text{if } 5 \le y_{ad+1} \le 9 \end{cases}$$

5.3 *Adaptive FLANN-Genetic (AFGA) Technique*

This approach, paid an impact on two different parameters for its advancement, i.e., probability for mutation (P_m) and probability for cross over (P_c). Updated values of $(P_m)_{k+1}$ and $(P_c)_{k+1}$ is calculated by using the following equations:

$$(P_m)_{k+1} = (P_m)_i - \frac{C_2 * n}{5} \tag{5}$$

$$(P_c)_{k+1} = (P_c)_i - \frac{C_1 * n}{5} \tag{6}$$

5.4 *FLANN Particle Swarm Optimization (FPSO) Technique*

It is based on the principle of particle swarm optimization and Function link neural network. PSO is utilized to update the weight at learning phase. Figure 5 represents the execution of PSO. Formula to calculate the fitness value is:

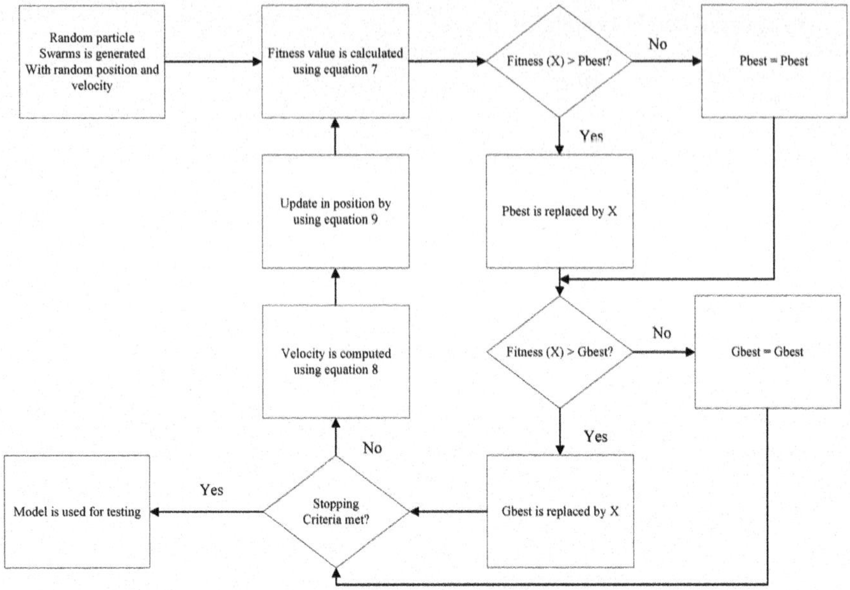

Fig. 5 Flowchart representing PSO execution

$$F_i = 1/E_i \tag{7}$$

$$V_{k+1}^i = V_k^i + C_1 * R_1 * (Pbest_k^i - X_k^i) + C_2 * R_2 * (Gbest_k^n - X_k^i) \tag{8}$$

$$X_{k+1}^i = X_k^i + V_{k+1}^i \tag{9}$$

where X is the position of particles and V is the velocity.

5.5 *FLANN-Clonal Selection Algorithm (FCSA) Approach*

FCSA is a hybrid approach using clonal selection algorithm and functional link neural network [13].

5.6 *Modified-FLANN Particle Swarm Optimization (MFPSO) Technique*

The main difference between PSO and MFPSO approach is that in case of MFPSO mutation stage is included just the completion of first stage. The following equation is required to calculate the update value of mutation.

$$(P_m)_{k+1} = (P_m)_i - \frac{C * n}{10} \tag{10}$$

where P_m is the first state of mutation and n is the generation number.

6 Experimental Setup

In this section of the chapter, we discuss the experimental setup done to find that developed malware detection model is effective or not. Six different hybrid functional link artificial neural network machine learning algorithms are implemented in this chapter. In Fig. 6, representation of proposed framework is demonstrated. In the first phase, feature selection techniques are implemented, i.e., PCA and rough set theory to select significant features. In the second phase, to normalize the features min-max approach is implemented. Distinct malware detections are developed by using six different machine learning techniques. After that, confusion matrix is developed by using the technique mentioned in [23, 24]. By comparing the malware detection model, best suitable model is selected and compared with the existing framework mentioned in the literature. If the detection rate is high after comparing the models with the existing framework, then proposed framework is useful or vice versa.

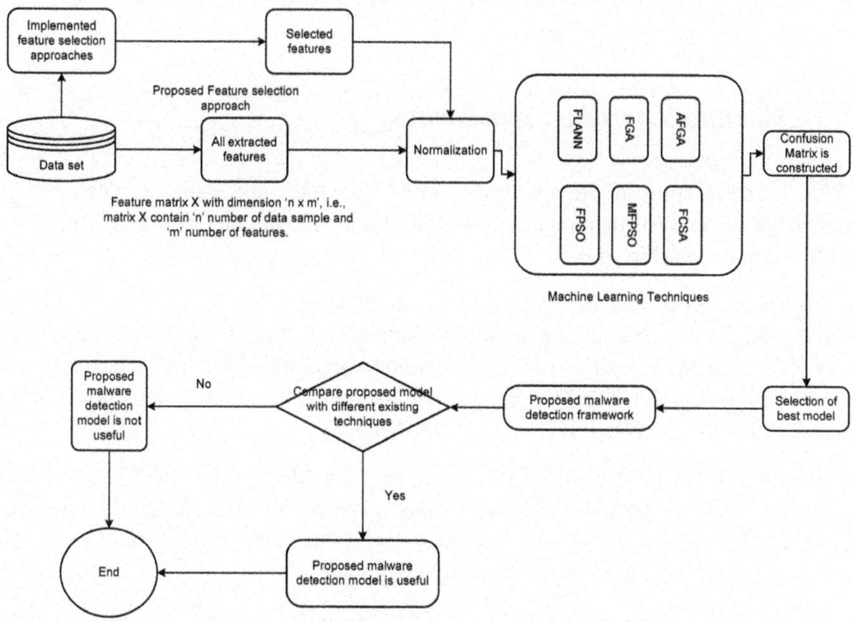

Fig. 6 Proposed framework, i.e., ANNDroid

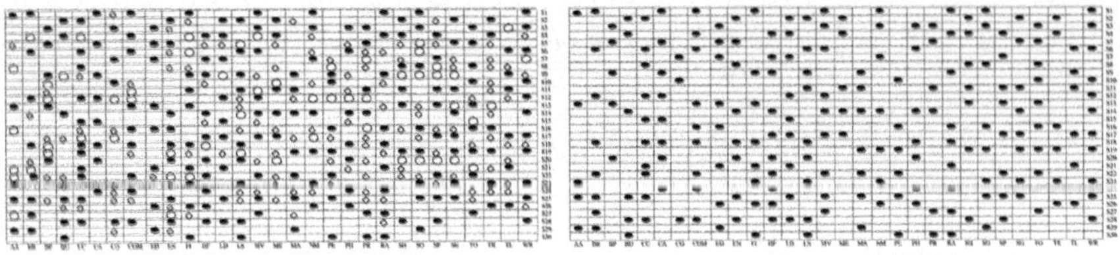

(a) PCA (b) RSA

Fig. 7 Feature selected using PCA and rough set analysis

7 Outcomes

In this section, the outcomes are gained by performing feature selection approaches and machine learning algorithms.

7.1 Feature Selection Approaches

Relevant features are selected using PCA whose eigenvalue is greater than 1, and features selected using rough set analysis are basis on heuristic search. Features selected using PCA and rough set analysis are demonstrated in Fig. 7.

7.2 Machine Learning Approaches

Tables 4 and 5 represent the measured value of accuracy and F-measure using PCA and rough set analysis using the equations mentioned in the literature [18, 19]. From tables, it may be inferred that:

- Highest detection rate is represented by bold value.
- It is observed from tables that models developed using features selection techniques achieved higher detection rate as compared to all extracted feature set.
- Model developed using FLANN-genetic accomplished higher detection rate as resembled to FLANN-PSO and FLANN-CSA.

In order to search, developed malware detection model is effective or not, box-plot diagrams of the individual developed model is constructed. Figures 8 and 9 demonstrate the box-plot diagrams for accuracy and F-measure using feature selection approaches. From figures, it can be concluded that:

- Based on Figs. 8 and 9, model developed by using RSA as feature selection technique achieve higher detection rate.

Table 4 Measured accuracy and F-measure using PCA

Accuracy												F-measure		
ID	AF	FLANN	FCSA	MFPSO	FPSO	AFGA	FGA	AF	FLANN	FCSA	MFPSO	FPSO	AFGA	FGA
DS1	62.33	73.33	75.0	77.37	77.66	**78.4**	0.68	**0.79**	0.77	0.76	**0.79**	0.72	0.71	**0.79**
DS2	68.18	72.66	78	76	73	76.6	**83**	0.63	0.73	0.72	0.76	0.77	0.76	**0.82**
DS3	70.8	84	83	82	82.6	81.6	**89.7**	0.72	0.84	0.83	0.82	0.84	0.83	**0.88**
DS4	62.8	76	75	72	75.6	79.6	**82**	0.68	0.70	0.72	0.79	0.78	0.77	**0.80**
DS5	70	82	85	81	86	87	**89**	0.67	0.86	0.84	0.82	0.84	0.81	**0.89**
DS6	68.8	78	76	81	86.6	85.6	**88**	0.71	0.72	0.76	0.80	0.82	0.83	**0.89**
DS7	67.8	87	87	85	80	88	**89**	0.71	0.84	0.82	0.86	0.88	0.89	**0.90**
DS8	67	83	84	85	88	87	**90**	0.71	0.82	0.82	0.83	0.83	0.84	**0.87**
DS9	78	90	90.8	**93**	91	91	92	0.78	0.90	0.92	**0.95**	0.90	0.90	0.94
DS10	66.8	87	86	85	89	88	**90**	0.68	0.77	0.75	0.82	0.84	0.83	**0.89**
DS11	69	88.7	88	81	85	87	**89**	0.68	0.84	0.83	0.83	0.84	0.83	**0.90**
DS12	70.8	87	85	88	82	85	**89**	0.71	0.84	0.83	0.86	0.86	0.88	**0.89**
DS13	71	86.7	87	88	88.2	88.1	**89.8**	0.67	0.81	0.85	0.83	0.84	0.83	**0.89**
DS14	77.3	85	87	86	87.6	87.3	**89.7**	0.71	0.88	0.83	0.84	0.83	0.86	**0.88**
DS15	70	89	89.9	89.8	91.8	90.7	**92.4**	0.73	0.84	0.85	0.88	0.88	0.89	**0.91**
DS16	67	83	82	80	84	85	**88**	0.70	0.86	0.84	0.85	0.84	0.83	**0.88**
DS17	79	87.7	88	88.4	87	86.8	**89.9**	0.76	0.85	0.86	0.88	0.86	0.88	**0.89**
DS18	79	88	87	88	86	89	**89.9**	0.69	0.83	0.85	0.86	0.86	0.86	**0.88**
DS19	68.9	82	83	84.8	85.8	86.9	**88**	0.72	0.84	0.81	0.82	0.80	0.82	**0.87**
DS20	68	80	81.9	88.3	**89.6**	81	89	0.68	0.78	0.79	0.80	**0.81**	0.80	0.79
DS21	67	87.8	88.7	89.8	90	89.7	**90.7**	0.74	0.84	0.84	0.83	0.81	0.80	**0.85**
DS22	70	86	81	82	84	85	**88**	0.70	0.83	0.84	0.83	0.84	0.85	**0.89**

(continued)

Table 4 (continued)

Accuracy												*F*-measure		
ID	AF	FLANN	FCSA	MFPSO	FPSO	AFGA	FGA	AF	FLANN	FCSA	MFPSO	FPSO	AFGA	FGA
DS23	68.9	88.78	83.71	83.9	**89**	84	88	0.66	0.81	**0.87**	0.83	0.82	0.80	0.81
DS24	65	88	88.2	**89.3**	89	87.7	82	0.70	0.84	0.81	**0.87**	0.82	0.85	0.85
DS25	78	88	89	87	89	84	**89.8**	0.71	0.85	0.83	0.82	0.81	0.80	**0.88**
DS26	66	80.8	82.1	84	85	85	**88**	0.72	0.85	0.84	0.85	0.86	0.86	**0.87**
DS27	67	88	81.9	89.6	87	86.9	**89**	0.67	0.81	0.82	0.83	0.84	0.82	**0.88**
DS28	67	89	88	**89.8**	86	81	88	0.71	0.86	0.84	0.81	0.86	0.85	**0.88**
DS29	67	81	87	**88.9**	85	84	81.9	0.67	0.87	0.86	0.87	**0.88**	0.81	0.82
DS30	60	84	89	89	88	81	**89.8**	0.67	0.84	0.85	0.86	0.83	0.81	**0.88**

AF stands for all extracted features

Table 5 Measured accuracy and F-measure using rough set analysis

Accuracy												F-measure		
ID	AF	FLANN	FCSA	MFPSO	FPSO	AFGA	FGA	AF	FLANN	FCSA	MFPSO	FPSO	AFGA	FGA
DS1	68.33	82	83	85	86	83	**89.8**	0.79	0.83	0.85	0.82	0.87	0.81	**0.89**
DS2	65	85	82	85	86	89	**91.8**	0.75	0.81	0.85	0.83	0.85	0.81	**0.87**
DS3	67	85	83	81	84	89	**90.8**	0.78	0.85	0.86	0.85	0.84	0.87	**0.89**
DS4	62.8	83	89	85	87	89	**90.7**	0.72	0.86	0.86	0.87	0.81	0.86	**0.88**
DS5	68.8	86	87	82	83	85	**89.8**	0.67	0.83	0.84	0.85	0.86	0.87	**0.90**
DS6	67.9	84	88	89	92	94	**96.7**	0.69	0.87	0.85	0.88	0.87	0.88	**0.90**
DS7	78	89.6	88.7	86	86.8	89.7	**93.8**	0.70	0.89	0.86	0.87	0.82	0.81	**0.89**
DS8	65	84	85	86	87	88	**91**	0.67	0.84	0.83	0.84	0.84	0.88	**0.89**
DS9	68	83	84	**96**	95	93	86	0.78	0.92	0.96	**0.99**	0.91	0.92	0.91
DS10	66.8	82	89	89	89.8	89.7	**97**	0.70	0.87	0.85	0.88	0.82	0.88	**0.96**
DS11	79	89	89	80	86	88	**98**	0.72	0.87	0.85	0.84	0.82	0.85	**0.93**
DS12	66.8	81	83	88	87	89	**90**	0.75	0.81	0.82	0.86	0.86	0.84	**0.89**
DS13	69.1	82	87	82	86	88	**89.8**	0.60	0.79	0.81	0.82	0.80	0.81	**0.88**
DS14	67	85	82	81	86	86	**90.9**	0.67	0.88	0.83	0.85	0.81	0.82	**0.89**
DS15	70.7	88	88	89.8	91	92	**97**	0.69	0.85	0.83	0.84	0.86	0.86	**0.94**
DS16	67	82	83	81	86	87	**88**	0.72	0.81	0.83	0.80	0.80	0.81	**0.83**
DS17	80	90	92	93	96	91	**98**	0.67	0.88	0.82	0.85	0.88	0.98	**1**
DS18	72	92.8	91	92.9	95	96	**98.9**	0.70	0.84	0.85	0.86	0.88	0.90	**0.93**
DS19	77	92	93	92	95	96	**97**	0.72	0.92	0.91	0.92	0.90	0.88	**0.95**
DS20	68	90	92	93	**95**	91	92	0.78	0.85	0.87	0.88	**0.89**	0.84	0.85
DS21	62	80	80.7	82	83	84	**85.7**	0.67	0.87	0.88	0.88	0.87	0.89	**0.9**

(continued)

Table 5 (continued)

Accuracy												F-measure		
ID	AF	FLANN	FCSA	MFPSO	FPSO	AFGA	FGA	AF	FLANN	FCSA	MFPSO	FPSO	AFGA	FGA
DS22	69.8	88	90	92	96	95	**98**	0.68	0.82	0.85	0.85	0.88	0.89	**0.91**
DS23	68.9	87.78	87.71	87.9	**91**	90	90.1	0.68	0.82	**0.85**	0.83	0.82	0.80	0.82
DS24	65	89	89.2	**91.3**	90	89.7	88	0.67	0.85	0.82	**0.89**	0.82	0.85	0.85
DS25	69.9	97	92	93	96	97	**98.8**	0.77	0.88	0.89	0.89	0.89	0.88	**1**
DS26	69.9	94.8	96.1	94	97	95	**97.9**	0.73	0.86	0.86	0.85	0.88	0.86	**0.92**
DS27	67	90.1	91.9	93.6	97	95.8	**97.9**	0.71	0.89	0.87	0.85	0.86	0.87	**0.91**
DS28	63	92	91	**98**	86	81	91	0.78	0.85	0.86	**0.89**	0.88	0.85	0.88
DS29	67	82	87	**92**	85	84	89	0.77	0.88	0.87	0.86	**0.87**	0.85	0.88
DS30	60	89	91	92	95	91	**98**	0.67	0.86	0.87	0.85	0.87	0.87	**1**

AF stands for all extracted features

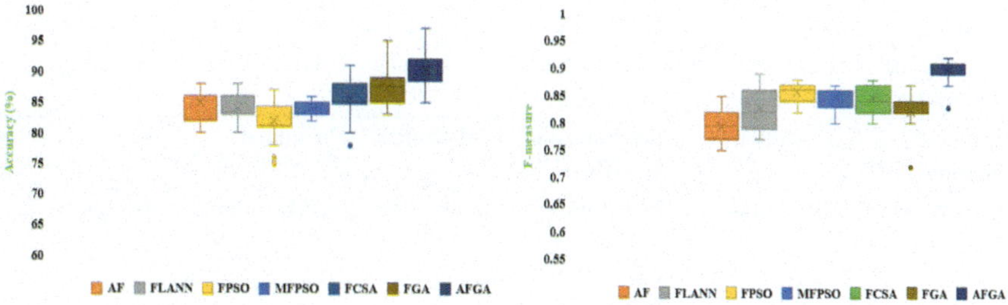

(a) Accuracy (b) F-Measure

Fig. 8 Box-plot diagram of accuracy and *F*-measure using PCA

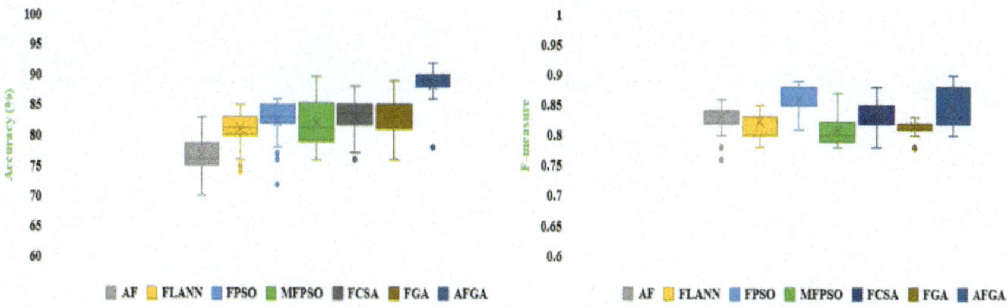

(a) Accuracy (b) F-Measure

Fig. 9 Box-plot diagram of accuracy and *F*-measure using RSA

- On the basis of Fig. 9, it is seen that model developed by using FLANN-genetic is having higher median value and few outliers. Model build by using RSA achieved higher detection rate as compared to PCA.

7.3 Comparison with Existing Developed Frameworks

In order to find out developed malware detection model is effective in detecting malware or not, in this chapter comparison is done by using existing frameworks present in the literature. To perform this experiment, freely available dataset; i.e., Drebin [2] is considered. Table 6 represent the comparison with existing approaches or frameworks presented in the literature.

7.3.1 Experimental Findings

In this chapter, a framework is developed by using android apps and by utilizing hybrid artificial neural network. Based on the outcome, this study is able to answer the questions discussed in Sect. 2.

Table 6 Comparison of developed model with available frameworks present in the literature

Framework/approach	Purpose	Approach	Deployment	Data set	Accuracy
Paranoid Android [27]	Detection	Dynamic and behavioral	Off-device	Limited	–
Crowdroid [3]	Detection	Dynamic, system call/API and behavioral	Distributed	Very-limited	High
Aurasium [33]	Detection	Dynamic and behavioral	Off-device	Limited	High
CopperDroid [29]	Analysis and detection	Dynamic, system/API and VMI	Off-device	Limited	Moderate
TaintDroid [5]	Detection	Run-time system call/API and behavioral	Off-device	Very-limited	Moderate
HinDroid [8]	Detection	Dynamic and API	Off-device	Limited	Moderate
Mahindru and Singh [25]	Detection	Run-time	Off-device	Limited	Moderate
MalDozer [10]	Detection	Run-time	Off-device	Limited	Moderate
DroidDet[36]	Detection	Static	Off-device	Limited	Moderate
Wei Wang[31]	Detection	Run-time	Off-device	Limited	Moderate
DeepDroid [14]	Detection	Run-time	Off-device	Limited	Moderate
PerbDroid [20]	Detection	Run-time	Off-device	Limited	High
Mahindru and Sangal [16]	Detection	Run-time	Off-device	Limited	High
ANNDroid (our proposed framework)	Detection	Run-time, permissions, API calls, user-rating and number of user download app	Off-device	Unlimited	Higher

RQ1: In the present study, implementation of six different machine learning techniques is used to develop malware detection model. Based on Tables 4 and 5, it can be implicit that model build using FLANN-genetic is more effective in detecting malware-infected from android.

RQ2: Yes, proposed detection model is effective in identifying malware-infected apps when compared to existing frameworks present in the literature.

RQ3: From Tables 4 and 5, it can be concluded that feature selection techniques have a significant role in building the malware detection model. Models developed using feature selection techniques are very effective when compared to the model developed using all extracted features.

8 Conclusion

This chapter paid a significant role while developing the malware detection models by using distinct android apps. In addition to that, it is observed that feature selection approach also paid an significant role while selecting the relevant features from all extracted features. Moreover, model developed using hybrid approach is more capable in detecting malware as compared to previously developed frameworks.

References

1. Arora, A., Peddoju, S.K., Conti, M.: Permpair: android malware detection using permission pairs. IEEE Trans. Inf. Forensics Secur. **15**, 1968–1982 (2019)
2. Arp, D., Spreitzenbarth, M., Hubner, M., Gascon, H., Rieck, K., Siemens, C.: Drebin: effective and explainable detection of android malware in your pocket. NDSS **14**, 23–26 (2014)
3. Burguera, I., Zurutuza, U., Nadjm-Tehrani, S.: Crowdroid: behavior-based malware detection system for android. In: Proceedings of the 1st ACM Workshop on Security and Privacy in Smartphones and Mobile Devices, pp. 15–26 (2011)
4. Cai, H., Meng, N., Ryder, B., Yao, D.: Droidcat: effective android malware detection and categorization via app-level profiling. IEEE Trans. Inf. Forensics Secur. **14**(6), 1455–1470 (2018)
5. Enck, W., Gilbert, P., Han, S., Tendulkar, V., Chun, B.G., Cox, L.P., Jung, J., McDaniel, P., Sheth, A.N.: Taintdroid: an information-flow tracking system for realtime privacy monitoring on smartphones. ACM Trans. Comput. Syst. (TOCS) **32**(2), 1–29 (2014)
6. Hou, S., Saas, A., Chen, L., Ye, Y.: Deep4maldroid: a deep learning framework for android malware detection based on linux kernel system call graphs. In: 2016 IEEE/WIC/ACM International Conference on Web Intelligence Workshops (WIW), pp. 104–111. IEEE (2016)
7. Hou, S., Saas, A., Ye, Y., Chen, L.: Droiddelver: an android malware detection system using deep belief network based on api call blocks. In: International Conference on Web-Age Information Management, pp. 54–66. Springer (2016)
8. Hou, S., Ye, Y., Song, Y., Abdulhayoglu, M.: Hindroid: an intelligent android malware detection system based on structured heterogeneous information network. In: Proceedings of the 23rd ACM SIGKDD International Conference on Knowledge Discovery and Data Mining, pp. 1507–1515 (2017)

9. Kadir, A.F.A., Stakhanova, N., Ghorbani, A.A.: Android botnets: What URLs are telling us. In: International Conference on Network and System Security, pp. 78–91. Springer (2015)
10. Karbab, E.B., Debbabi, M., Derhab, A., Mouheb, D.: Maldozer: automatic framework for android malware detection using deep learning. Digit. Invest. **24**, S48–S59 (2018)
11. Kim, T., Kang, B., Rho, M., Sezer, S., Im, E.G.: A multimodal deep learning method for android malware detection using various features. IEEE Trans. Inf. Forensics Secur. **14**(3), 773–788 (2018)
12. Kumar, L., Hota, C., Mahindru, A., Neti, L.B.M.: Android malware prediction using extreme learning machine with different kernel functions. In: Proceedings of the Asian Internet Engineering Conference, pp. 33–40 (2019)
13. Kumar, L., Rath, S.K.: Hybrid functional link artificial neural network approach for predicting maintainability of object-oriented software. J. Syst. Softw. **121**, 170–190 (2016)
14. Mahindru, A., Sangal, A.: Deepdroid: feature selection approach to detect android malware using deep learning. In: 2019 IEEE 10th International Conference on Software Engineering and Service Science (ICSESS), pp. 16–19. IEEE (2019)
15. Mahindru, A., Sangal, A.: Dldroid: feature selection based malware detection framework for android apps developed during covid-19. Int. J. Emerg. Technol. **11**(3), 516–525 (2020)
16. Mahindru, A., Sangal, A.: Feature-based semi-supervised learning to detect malware from android. In: Automated Software Engineering: A Deep Learning-Based Approach, pp. 93–118. Springer (2020)
17. Mahindru, A., Sangal, A.: Gadroid: a framework for malware detection from android by using genetic algorithm as feature selection approach. Int. J. Adv. Sci. Technol. **29**(5), 5532–5543 (2020)
18. Mahindru, A., Sangal, A.: Mldroid-framework for android malware detection using machine learning techniques. Neural Comput. Appl., 1–58 (2020)
19. Mahindru, A., Sangal, A.: Parudroid: validation of android malware detection dataset. J. Cybersecur. Inform. Manag. **3**(2), 42–52 (2020)
20. Mahindru, A., Sangal, A.: Perbdroid: effective malware detection model developed using machine learning classification techniques. In: A Journey Towards Bio-Inspired Techniques in Software Engineering, pp. 103–139. Springer (2020)
21. Mahindru, A., Sangal, A.: Semidroid: a behavioral malware detector based on unsupervised machine learning techniques using feature selection approaches. Int. J. Mach. Learn. Cybernet., 1–43 (2020)
22. Mahindru, A., Sangal, A.: Somdroid: android malware detection by artificial neural network trained using unsupervised learning. Evol. Intell., 1–31 (2020)
23. Mahindru, A., Sangal, A.: Fsdroid:-a feature selection technique to detect malware from android using machine learning techniques. Multimedia Tools Appl., 1–53 (2021)
24. Mahindru, A., Sangal, A.: Hybridroid: an empirical analysis on effective malware detection model developed using ensemble methods. J. Supercomput., 1–43 (2021)
25. Mahindru, A., Singh, P.: Dynamic permissions based android malware detection using machine learning techniques. In: Proceedings of the 10th Innovations in Software Engineering Conference, pp. 202–210 (2017)
26. Martín, A., Menéndez, H.D., Camacho, D.: Mocdroid: multi-objective evolutionary classifier for android malware detection. Soft Comput. **21**(24), 7405–7415 (2017)
27. Portokalidis, G., Homburg, P., Anagnostakis, K., Bos, H.: Paranoid android: versatile protection for smartphones. In: Proceedings of the 26th Annual Computer Security Applications Conference, pp. 347–356 (2010)
28. Shen, F., Del Vecchio, J., Mohaisen, A., Ko, S.Y., Ziarek, L.: Android malware detection using complex-flows. IEEE Trans. Mob. Comput. **18**(6), 1231–1245 (2018)
29. Tam, K., Khan, S.J., Fattori, A., Cavallaro, L.: Copperdroid: Automatic reconstruction of android malware behaviors. In: NDSS (2015)
30. Tong, F., Yan, Z.: A hybrid approach of mobile malware detection in android. J. Parallel Distrib. Comput. **103**, 22–31 (2017)

31. Wang, W., Zhao, M., Wang, J.: Effective android malware detection with a hybrid model based on deep autoencoder and convolutional neural network. J. Ambient Intell. Hum. Comput. **10**(8), 3035–3043 (2019)
32. Xiao, X., Zhang, S., Mercaldo, F., Hu, G., Sangaiah, A.K.: Android malware detection based on system call sequences and LSTM. Multimedia Tools Appl. **78**(4), 3979–3999 (2019)
33. Xu, R., Saïdi, H., Anderson, R.: Aurasium: Practical policy enforcement for android applications. In: Presented as Part of the 21st USENIX Security Symposium (USENIX Security 12), pp. 539–552 (2012)
34. Yerima, S.Y., Sezer, S.: Droidfusion: a novel multilevel classifier fusion approach for android malware detection. IEEE Trans. Cybernet. **49**(2), 453–466 (2018)
35. Zhou, Y., Jiang, X.: Dissecting android malware: characterization and evolution. In: 2012 IEEE Symposium on Security and Privacy, pp. 95–109. IEEE (2012)
36. Zhu, H.J., You, Z.H., Zhu, Z.X., Shi, W.L., Chen, X., Cheng, L.: Droiddet: effective and robust detection of android malware using static analysis along with rotation forest model. Neurocomputing **272**, 638–646 (2018)

DanVeer: A Secure Resource Funding Mobile Application

Himesh Nayak, Rahul Johari, and Haresh Nayak

1 Introduction

Mobile application development is the process of developing software applications designed to run on mobile devices, such as a smartphone or tablet computer. The beginning of mobile app development dates back to 1993. Since then, mobile app development has seen many advancements, and there are many new technologies also coming up developing mobile applications. But still, there is no standard procedure to build a mobile application. In this paper, we talk about a resource funding application built with android native Java and implement the concepts of blockchain. We have reviewed the application in this paper and have discussed the various techniques that are used in the development of an app starting from the architecture of the application to the security and privacy of the users' data.

H. Nayak · R. Johari (✉)
SWINGER: Security, Wireless, IoT Network Group of Engineering and Research, Guru Gobind Singh Indraprastha University, Sector-16C, Dwarka, Delhi 110078, India
e-mail: rahul@ipu.ac.in

H. Nayak
Delhi Technological University, Bawana Road, New Delhi, Delhi 110042, India

J. Singh et al. (eds.), *Mobile Application Development: Practice and Experience*, Studies in Systems, Decision and Control 452,
https://doi.org/10.1007/978-981-19-6893-8_6

2 Motivation

After the pandemic, as the world is changing its lifestyle, there has been no significant change in the methods of seeking materialistic help. Though social media has emerged as an instant star for users, the process is still tiring. Moreover, the lack of knowledge in rural areas can lead to some needy left hanging even if the resource is available.

Thus, we have tried to build a digital architecture in the form of the mobile app, DanVeer, to ensure that no one faces such problems just because they do not have enough knowledge. Since such applications contain some crucial data in it which can be attacked by a lot of hackers and black marketeers, we have tried to build a secure application that can be accessible to most of the users.

3 Literature Survey

- Google [1] has provided official docs which tell us about some of the standard procedures to develop an android application. Starting from building a single page application, to a complex application with a proper structure, they have mentioned it in the docs.
- Thomas and Devi [2] in their research have given an overview of the Mobile Application Industry and have talked about its predominance in India. They have discussed the mobile app development cycle and the various procedures involved in the development of an application.
- Patidar and Suman [3] have presented a review of various available mobile app characteristics concerning present mobile apps. They have discussed the selection of mobile app paradigms and the utilization of the mobile app characteristics.
- Liu and Shestak [4] have discussed that security and privacy issues which currently are the burning issues and therefore are at the forefront of mobile app development. They proposed an approach for understanding how mobile crowdfunding app producers should ensure the privacy and responsibility of each member of the "crowd" involved in any crowdfunding campaign.
- Hayes et al. [5] in their research paper have raised several concerns about the collection and sharing of personal data conducted by mobile apps without the knowledge or consent of the user. They have demonstrated that permissions and privacy policies are not enough to determine how invasive an app is.
- Braham et al. [6] have examined the role of design patterns and ontology models in order to help with the generation of mobile applications, which can be adapted at run time to the various user needs, different context scenarios, interactive design modes or technology requirements.

4 About the DanVeer Application

DanVeer is an app that can become a go-to solution for the people who are looking for resources, in the time of a natural disaster or a pandemic and the people and organizations who are willing to provide the resources. These resources can be of any kind, such as monetary, food, clothes, etc. Since money transactions and the details of the users are too sensitive to be shared publicly, DanVeer is backed by cryptography algorithms and secure technologies like cryptocurrency and blockchain. The users will be authenticated before entering the app so that there is no breach of data. They can register as an individual or an organization by providing the required details.

Once the user is logged into the app, they can view the requests made by other users and respond to them monetarily or by providing the details of the resource they possess. If they choose to provide monetary assistance to the requester, they can simply use the inbuilt payment gateway. Another level of security is added before the transaction by using ReCAPTCHA to check if there are any bots/auto-run programs making the transaction.

If they choose to provide the details of the resource, they can enter the required details in the app, and it will be sent to the requester. The requester can then view the information and act accordingly. All these transactions will be encrypted by using secure cryptographic algorithms such as SHA-256 and data encryption system so that no infringement can take place. It will also prevent hackers and those with malicious intent to access the request data.

Thus, DanVeer acts as a secure medium, removing the middleman, to fulfilling the requirements of an individual in times of need. When used with pure intentions, it can surely prove to be a boon for society.

5 Flowchart of the Application

The flowchart in Fig. 1 shows the exact working of the application. It depicts the various activities that can be done through the application which are adding a request for a particular resource and accepting a request. Once the application is started, the authenticity of the user is checked. If the user is logged in, the control of the application is sent to the Home Page, else the login/register form appears on the screen. Further in the flowchart, we can see that when the control reaches the Home Screen, the user can either make a request or view a request. Data is fetched from the database, and when a particular transaction is completed, the data is encrypted and sent to the database.

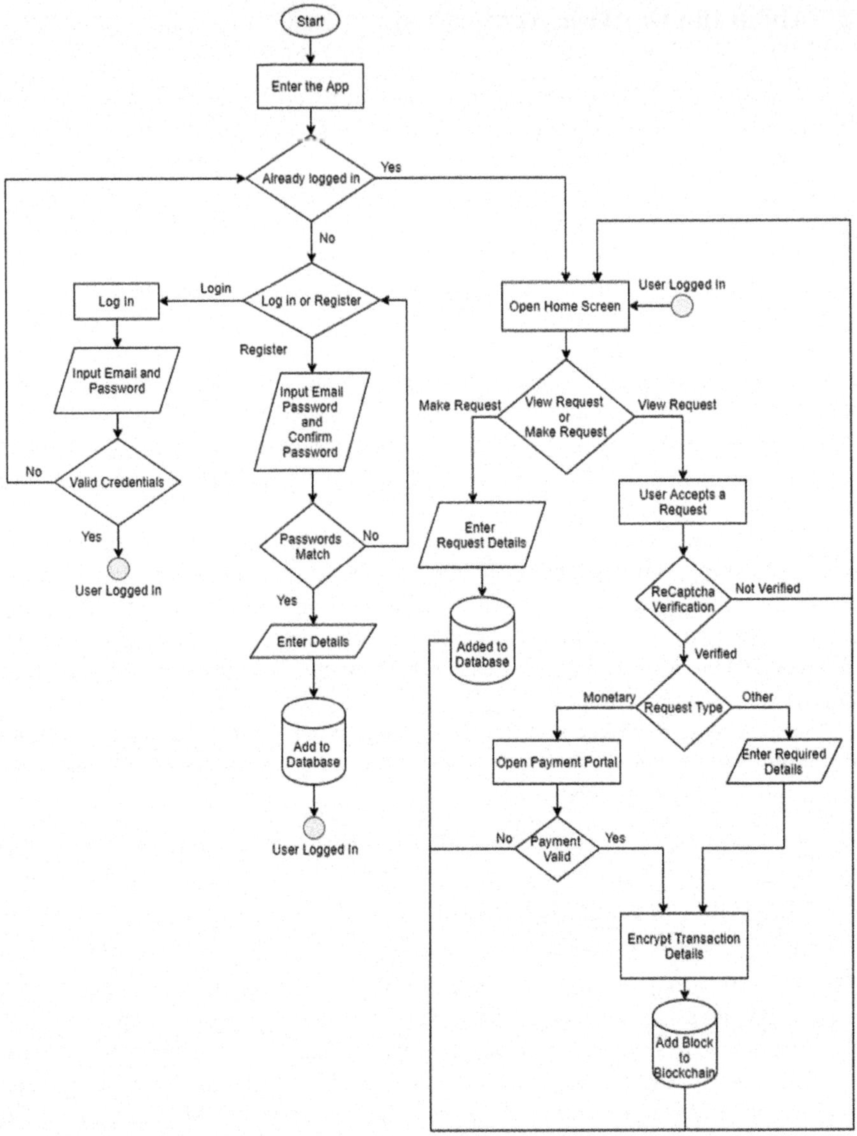

Fig. 1 Flowchart of the working application

6 Algorithm

Algorithm 1 Accepting a Monetary Request

1. **Start**.
2. Open App

3. **Start** "*Authentication*":
4. **If**(User is Logged In):
5. Fetch from Database "*User Data*"
6. Open Screen "*Home Screen*"
7. **Else**:
8. Open Screen "*Login/Register*"
9. User Taps Button "*Login*"
10. Open Card "*Enter Credentials*"
11. User Enters "*Email and Password*"
12. **If** (Valid Credentials):
13. Log In "*User*"
14. **Else**:
15. **End** "*Authentication*"
16. **Return** "*Step 3*"
17. **End If**.
18. Fetch from Database "*User Data*"
19. Open Screen "*Home Screen*"
20. **End If**.
21. **End** "*Authentication*"
22. **Start** "*Accept Monetary Request*":
23. Open Screen "*Monetary Requests*"
24. User Selects "*Request*"
25. Open Screen "*View More Screen*"
26. User Selects "*Pay*"
27. Open Card "*ReCAPTCHA*"
28. **If** (Valid ReCAPTCHA):
29. Open Card "*Payment Portal*"
30. User Selects "*Payment Method*"
31. **If** (Payment Successful):
32. Build "*New Block*"
33. Add Block to "*Blockchain*"
34. **Else**:
35. **Return** "*Step 25*"
36. **End If**.
37. **Else**:
38. **Return** "*Step 25*"
39. **End If**.
40. **End** "*Accept Monetary Request*"
41. **Stop**.

The algorithm depicts the steps followed for accepting a monetary request, from the starting of the application to accepting the request. It shows, how the application checks for bots using ReCAPTCHA to increase security, and when the monetary transaction is completed, a new block is created and added to the blockchain.

Fig. 2 These are the screenshots of DanVeer application

7 App Screenshots

The screenshots of the DanVeer app are depicted in Fig. 2.

8 Application Code

The complete code of the app can be found on the GitHub [7].

9 Architecture for Generating Applications

There are many applications in the market that are developed without using a proper architecture, but there are many useful architectures that can be used to develop dependable applications which can cater to the needs of the clients. Some of the architectures that are present for mobile application development are as follows.

9.1 Model View Controller (MVC) and Model View Presenter (MVP)

The model view controller divides the application into three main logical parts, namely the model, the view and the controller. Each of these components handles specific development aspects of the application such as view: the UI part, controller: business logic and business rules, and model: database connectivity part.

Model–View–Presenter is a user interface architectural pattern engineered to facilitate automated unit testing and improve the separation of concerns in presentation logic.

The three components of MVP are:

- Model: It is the layer for storing data.
- View: It represents the UI logic layer.
- Presenter: It decides what to display by applying the data from Model to Views.

MVP is widely accepted due to its modularity, testability and a cleaner and more maintainable codebase.

9.2 Model View ViewModel (MVVM)

Model View ViewModel is a software architectural pattern that is used to separate the development of the graphical user interface (also known as the view) from the development of the business logic or back-end logic (also known as the model) to make the view independent of any specific model platform. DanVeer is based on the MVVM architecture which lets us use the live data from the database and show it in the app.

10 App Debugging and Testing

When developing any application, the developer should debug the code in order to develop the application better. Android Studio is the official integrated development

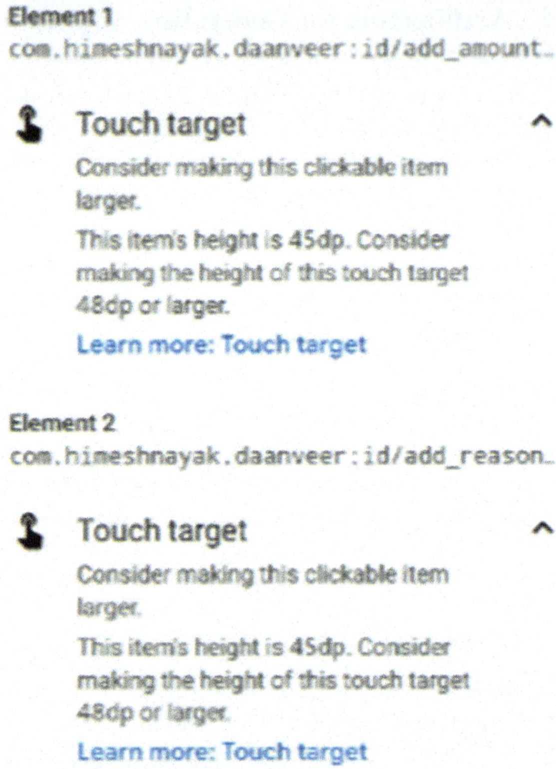

Fig. 3 Accessibility test results of DanVeer app for elements 1 and 2

environment (IDE) for Google's android operating system, and it provides a debugger to debug the application.

Testing of any code is most important as it tells us about the ways we can improve the code. Various aspects of an application need to be checked before deploying it and sharing it with the users. The accessibility test was by the scanner application [8] on a previous version of the app, and it tells us a lot of detailed user interface errors that could be improved in order to make the application better.

The results shown in Figs. 3 and 4 clearly show that the prototype of the application has a few accessibility issues which can be corrected to make the application better for the users. The suggestions given in the result are mostly related to the color contrast of the different parts in the application.

11 App Review and Analytics

Once the app has been developed and shared with the users, it is very important to review the usage of the app and measure the usage of the app. A developer should

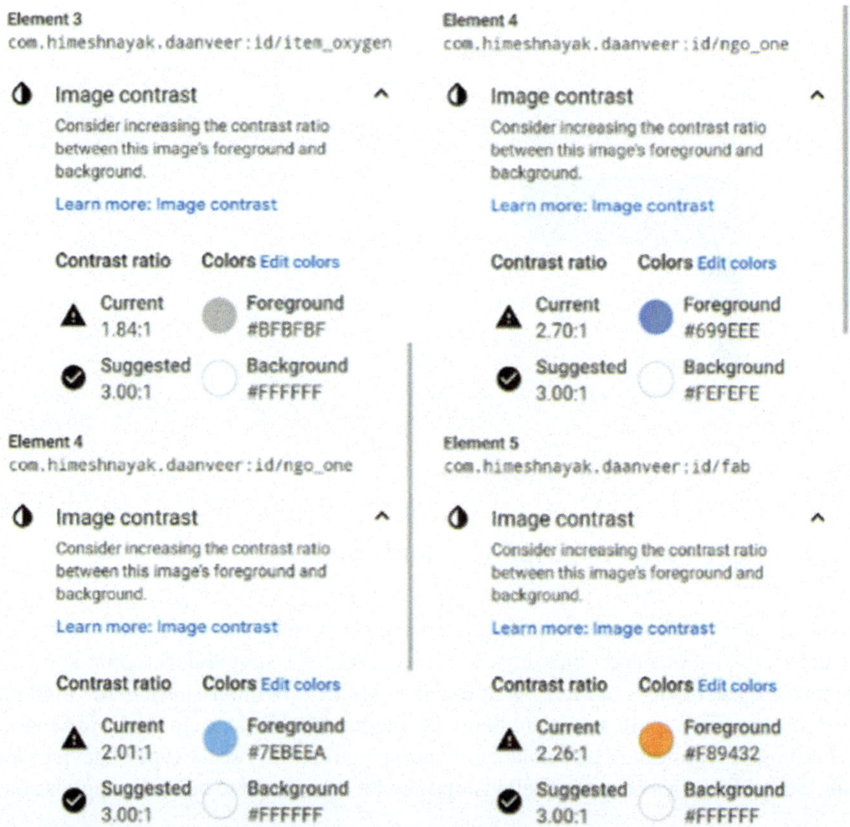

Fig. 4 Accessibility test results of DanVeer app for elements 3 and 4

be aware of the problems the users are facing while using the application and which features of the application are being used the most by the users. Google Analytics as shown in Fig. 5 helps one understand how people use the Web, Apple or android app.

DanVeer is an application where the users can ask out for the resources which they need. Thus, we should know which type of resource is needed by the users, so as to add some feature to the application to highlight that resource and make it more visible to the people who are donating. For achieving this, we can add the Google Analytics integrated with Firebase [9] and record the per page user engagement for the various resources such as oxygen cylinders, food, etc.

The above figure depicts the usage of the various application screens. It shows graphically which screen is opened the most that can tell us which resource is required the most.

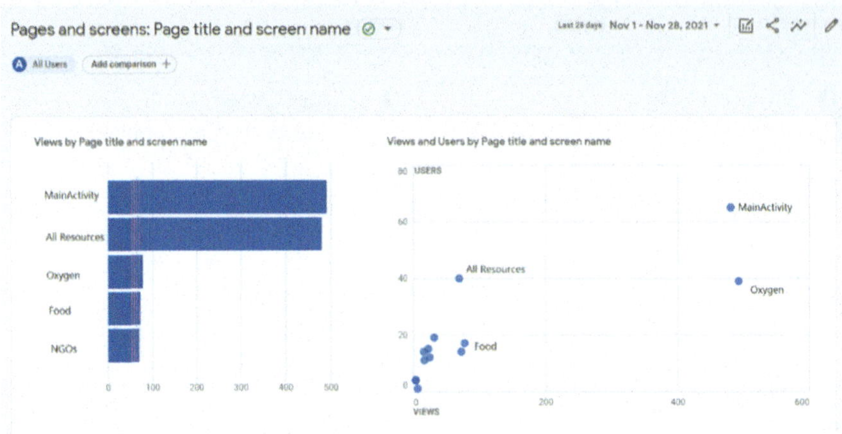

Fig. 5 Google analytics of DanVeer application

11.1 App Beyond Smartphones and Tablets

Apps are not just bound to mobile phones anymore. With the emergence of smartwatches, televisions and other devices, developers have started developing applications for these devices as well. Wearable devices are now flourishing in the market, and thus, various applications are being developed for the same. In Android Studio, when you start building a particular application, you can choose the type of device like any wearable device or a smart television for which you want to build the application.

11.2 App Migration Across Platforms

Nowadays, cross-platform applications are also into fashion and are being used by many developers around the world. One such upcoming cross-platform framework is Flutter which can build applications for both android and iOS platforms. DanVeer is an android native application that is developed in Java and currently is available for android users. But if a clone of the application is developed using flutter, the same source code can be then compiled for both android and iOS.

11.3 Maintenance of Apps

The maintenance of an application is essential for any developer. With the increase of users and technologies, one has to keep updating the application to cater to different users and devices. Most app development companies have a large number of developers working on the application, and thus, it becomes easier for them to maintain the application, whereas there are some applications that are open-sourced. With the number of developers increasing in the world day by day, the open-source community is also increasing and the people can report a bug and request solutions for it in the community.

11.4 App Security and Privacy

A lot of data is taken in the apps. Google has been planning to release android's new version, "Android 12", which has a lot of features for the protection of the user's privacy. These features can not only help the user to keep his sensitive data safe but also will warn the users of suspicious activities done by some apps that might compromise the privacy. Some of the features announced are as follows. The app is being checked with some testing strategies [10]. It was found to be accomplish all such type of errors fixing.

11.4.1 Privacy Dashboard

Privacy Dashboard lets the user see how different apps are accessing data on his device. They can see any specific type of feature (such as "Location" or "Camera") and get a timeline that tells, exactly which apps have been accessing that feature when. This feature will help the users to keep a check on the suspicious activities of some apps.

11.4.2 Toggle Sensor Switches

A new series of toggle switches are presented that let the users turn their phone's camera, microphone or GPS sensor completely off with a single fast tap. This feature can help the users in some situations where they do not want any specific application to use that particular service.

12 Conclusion

We are living in the era of mobile applications and depend upon them for almost every task from booking a cab to ordering food. There should be a standard procedure and testing techniques that can help build effective and useable mobile applications. We have tried to research the various guidelines given for mobile application development and developed an application to solve the problems faced by society. There are still some new techniques and different tests that can be done on the application to improve it so that it can be helpful for a large part of society.

References

1. Google: Android Development Documentation (2021). https://developer.android.com/docs/. [Online; accessed 21-November-2021]
2. Thomas, C.G., Devi, A.J.: A study and overview of the mobile app development industry (2021)
3. Patidar, A., Suman, U.: Towards analysing mobile app characteristics for mobile software development. In: 2021 8th International Conference on Computing for Sustainable Global Development (INDIACom) (2021)
4. Liu, Z., Shestak, V.: Issues of crowdsourcing and mobile app development through the intellectual property protection of third parties (2021)
5. Hayes, D., Cappa, F., Le-Khac, N.A.: An effective approach to mobile device management: security and privacy issues associated with mobile applications. Digital Bus. **1**(1), 100001 (2020)
6. Braham, A., Buendía, F., Khemaja, M., Gargouri, F.: User interface design patterns and ontology models for adaptive mobile applications (2020)
7. Nayak, H.: DanVeer application code on GitHub (2021). https://github.com/HimeshNayak/Daan-Veer [Online; accessed 20-November-2021]
8. Google: Scanner App on Google Play Store (2021). https://play.google.com/store/apps/details?id=com.google.android.apps.accessibility.auditor. [Online; accessed 20-November-2021]
9. Google: Google Analytics and Firebase Documentation (2021). https://firebase.google.com/docs/analytics/. [Online; accessed 21-November-2021]
10. Ghosh, D., Singh, J.: Effective spectrum-based technique for software fault finding. Int. J. Inf. Technol. **12**(3), 677–682 (2020)

Mobile Data Analytics: A Comprehensive Case Study

Akash Bhattacharyya and Jagannath Singh

1 Introduction

Mobile applications are software which are dedicated to be used on smartphones, tablets and different handheld devices. These mobile applications or apps are developed with the help of programming languages which dictate the primary working of the mobile platform it is intended for: ObjectiveC is used for iOS, Java is used for Android, C# is used for Windows phones, etc. [1]. Similarly each type of apps is distributed by a specific distribution channel, such as App Store for iOS apps and Google Play for Android apps. The distribution channels offers around one million different types of apps to be downloaded. Islam et al. [1] have shared the statistics that the development of these apps are having a direct impact on the economic growth as well as the social perspective. The apps business has been found to generate an estimated revenue of $4.5 billion USD in the year 2009.

Mobile data analytics has been found to be an area of growing relevance and interest for the IT professionals and academicians [2]. This has been found to rise due to the high rise in the number of mobile device users and their affinity to complete their work on their mobile devices. It can be seen that almost every individual on Earth has at least one mobile device which they are using to complete their work. Moreover with recent effect of pandemic and severe lockdown, mechanisms ordered by the governments around the world have forced the individuals to work on their mobile devices from home [3]. Thus, it can be said that the use of mobile data analytics can be found to be of the highest priority in order to provide the users with the best experience so that they can complete their work in time and also stay back with the application provider.

A. Bhattacharyya (✉) · J. Singh
School of Computer Engineering, KIIT Deemed to be University, Bhubaneswar, Odisha, India
e-mail: 2081008@kiit.ac.in

J. Singh et al. (eds.), *Mobile Application Development: Practice and Experience*,
Studies in Systems, Decision and Control 452,
https://doi.org/10.1007/978-981-19-6893-8_7

2 Mobile Analytics

Mobile analytics makes use of data generated by mobile sites and applications in order to measure and analyze them. Mobile analytics has recently emerged with the rise in mobile data and the corresponding advances in the field of mobile computing [4]. However in the field of mobile data analytics, the challenges which are encountered are from the data noise, location awareness and data redundancy of the mobile data collected. To provide the users with quality satisfaction, the use of mobile data analytics should be able to provide real-time intelligent decision-making system. Due to the increase in the use of knowledge discovery, the development of complex mobile system to facilitate the process of mobile analytics with the help of empirical data has become easier to implement.

AT Internet analytics solution helps in the process of tracking and measuring the way the mobile users are interacting with the mobiles sites and applications [5]. With the help of such mobile data analytics processing, one can improve the process of cross-channel marketing and eventually optimize the movie use experience for the customers. The analytics would also help in improving the user engagement time on the application as well as user retention. However at the end of the day, the user may convert to another application due to the availability of a specific feature which he has found useful than the one he was currently using [6]. Single channel of data analysis of the data being collect may not be enough to reach the conclusion required for the company. It would be beneficial to measure the lifetime data collected which would help in the process of the analysis of the behavior than measuring the isolation channels of data. In order to gain the highest effect of the analytics from the mobile data being collected, it would be beneficial to process it against data collected from other channels such as email, web and various social media networks which can help in understanding the mobile app usage [7]. This format of data analysis can be helpful even if the user does not convert to another application or makes any kind in-app purchases.

3 Importance

The rise of use of smartphones was not greatly welcomed by the consumer forum as they were not able to complete large forms of tasks on the smartphones at that time. With the current development of mobile operating systems by Android and iOS and the availability of large number of applications which can be used to complete any form of tasks which would have previously required the use of a personal computer or a laptop, it can be seen that the mobile application industry is the largest growing sector in today's economic world [8]. Any new organizations that are going public in the recent times or the companies who have been ruling the market for a long time, all had to provide the consumers with the required mobile application or mobile compatible websites for their services and products so that they would be able to retain their customer base. Companies have been found to specifically focus on the development and improvement of their mobile compatible applications and websites as the majority of the users in today's time try to complete their work on their personal mobile devices [9]. Thus, in order to provide good quality customer satisfaction, the developers needs to understand the requirement which is being requested by their customer base from the application or website. Keeping the customer engaged with the application and providing them with proper updates from time to time will help the developers to keep their customer base strong and happy.

4 Comparison with Web Analytics

During the rise of mobile applications, developers used to treat mobile application and desktop software as separate entity and use to control the usage statistics and analytics separately [10]. With the improvement in the quality of smartphones being made and the increase of processing speed of the mobile devices, the developers have started to treat both the sections as a single entity. The major difference in the two forms of devices is the screen size and the requirement of less number of clicks to complete a specific task. Moreover, it has been seen that the use of mobile phone in comparison to a desktop setup has been found to be faster and easier to complete [11]. However, issues remain where some websites have not been upgraded to be compatible with mobile screen sizes and thus becomes difficult to use on a mobile device.

5 Metrics Considered for Mobile Application Versus Website

Once a company starts to adapt the use of an Omni-channel strategy, it has been found that there is an absence of a specific metric for the evaluation of mobile application from a website. A simple process of tracking success in a mobile environment is through the use of events which is similar to page views in web [12]. Events are considered as actions which are done by a user on the mobile app such as launch, or clicking a specific button. With the help of event tracking, the analyst will be able to measure any form of behavioral analytics required by the stakeholders of the application.

The product managers can track all forms of event raised by an application in order to understand their customers. The daily active users can be measured with the help of frequency a customer engages with the application [13]. The activeness of the application can also be measure with the help of tracking whether the customer has push notification turned on or not or by the number of times the application is opened by the customer in a single day.

Each session the user engages with the application has help in the understanding of the engagement of the application with the users [14]. Another set of metrics can be related to the measurement of the revenue generated by the users like average revenue generated per customer or the average amount of value transaction completed.

Computation of retention analysis for a mobile application is bit different to measure. The basic process of calculation of retention process is done with the help of counting the number of customers who logged in to the application today and then measure the number of customers who reclogged in to the application a month later [15]. On the basis of retention analysis, the analyst will be able to understand how many times a user engages with the application and the amount of time the user spends on the application.

Apart from all these metric for the analysis of a mobile application, there are some performance metrics which are used to count the app crashes. The success metrics for a business project should be global and not specific to a mobile application.

6 Case Studies Suggesting Benefit of Mobile Analytics

6.1 Game Case Study: Pokémon Go

Pokémon GO is an augmented reality-based game developed by the joined venture of Niantic Labs and Nintendo. It was released on July 6th of 2016 for both Android and iOS platform. It makes use of user location to spawn an animated monster which can caught by the user and then trained in-game to fight battles against other players on the same platform [16]. The developers made use of real world locations to make poke stops and poke gyms as known in-game which is enabled to give out various items and other consumables required during the progress of the game.

It was one of the most downloaded games upon release with a statistics of 500 million by the end of 2016. The current statistics of the game amounted to 147 million daily user interactions by the end of May 2018 and thus was able to gross more than $6 billion in revenue by the end of 2020. The original animated monsters by the name of Pokémon was released long ago in the form of animated television series in the year 1996 [17]. This caused an insane rise in popularity of the game among people of all age who had grown up watching the same animation series. This Pokémon-based game was able to break all forms of record in the video game industry even breaking record of the games released under the same genre of Pokémon.

The origin story of Pokémon Go starts with the use of real world locations being used in another game developed by Niantic in the year 2013 named Ingress [18]. However, the game was not as popular as Pokémon Go, but user of both the game was able to spot out some of the common similarities between the two games.

The process of data collection from the users was done since the release of the game. The developers were first collecting data such as location designated as latitude and longitude and timestamps of the location. The data was being collected at every hour. The first issue occurred when the data was being collected and shared even when the game was not being played by the users. The Niantic developers stated that the 24 h data collection was a "bug" and has since been removed from the code. Developer's added incentive-based bonus items in the game when the users were able to visit some of the advertised locations which they received as quests completion in the game. The users are forced to locate virtual elements in the real world though the use of games interface and get incentive items. These type of quests in turn helped in the promotion of the advertised locations easily.

The second phase of data collection was found where the number of poke stops were relatively low in remote areas [19]. This was removed with the help of in game application-based suggestion system where players would be able to submit a location which can be converted into a poke stop or a poke gym. This helped in the growth of players in the location and boosted the spawn of Pokémon's. Moreover, the number of Pokémon spawn in an area is governed based on the number of user interaction the poke stops and poke gyms receive throughout the day.

6.2 Utility Case Study: Gboard

Gboard is a keyboard application developed by google for mobile devices. It was first release for iOS mobile devices and later on Google integrated the keyboard into their Android operating system. Partially being forced upon the users, the users were able to adapt to the use of the keyboard features rather easily [20]. The key features of the application were swipe typing where the user would be able to glide their finger over the keyboard starting from an alphabet and trailing over the subsequent alphabets of the word which made the keyboard to predict the current word which the user wanted to type. This form of typing was fast paced and quickly became a hit among the users of the keyboard. The next feature was the ability to use voice input which would be converted into text and written into the text box area. This has been improved over the years to make the algorithm understand the tone of voice languages the users from different areas of the world.

The main data collection process of the Gboard was concerned with the way a user types a sentence or chats on different applications. Google cleared the issue that the data being collected by them from the users were anonymous in nature which was helpful. Google also said that the data was being used to learn to create instant replies which has gained new perspective in the use of mobile devices to reply back to messages and emails easily [20]. Another form of data collection and analysis helps in the development of auto-correction process that has been incorporated in the keyboard. The latest features which has been rolled out based on this data analysis is the suggestion of emoji's during chatting on any messaging applications. The developers have also included the ability to create stickers and share gif images through the keyboard with the inbuilt ability to search the web and share relevant information.

The artificial intelligence associated with the Gboard can be said to be evolving on a daily basis and provided helpful ways to process data and type faster on a mobile device [18]. With the rise in the use of mobile devices around the world and the availability of email, chatting and social media integration applications within the mobile devices, the use of a robust keyboard which would be able to provide the users with the required amount of help in order to complete their work easily without switching to different applications has become the primary target of Gboard.

7 Existing Tools

This section discusses about some of the popularly used mobile data analytics tools which have been built to process the data collected from the applications to create analytics which would help in improving the performance of the mobile websites and applications [1]. The following table summarizes the details of some of the mobile analytics software.

Sl. No.	Name	Description	Cost
1	Localytics	This software is able to track at the rate of 100,000 monthly active users along with 12 million data points from the software. The package helps in keeping a track on the user flows and funnels as well as granular analytics from the data being collected. The software also helps in accessing the API of the software	Free up to 10,000 users Cost ranges from $200 to $1200 per month depending on the number of users
2	Google Analytics	Provides the users with integrated google products such as AdSense, AdWords and various others. Features includes data collection through Internet of things devices, configurable API, websites and mobile applications	Free and subscription-based usage on the amount of usage
3	Heap	It is able to collect data and analyze every action (click, swipe) made by the user automatically. No extra amount of coding is required by the user to view the analytics	Free usage along with subscription-based profiles catered to the requirement of the user
4	Kochava	It provides largest mobile data marketplace along with real-time data analysis	Free version and $100 per month subscription basis

(continued)

(continued)

Sl. No.	Name	Description	Cost
5	Mixpanel	A simple interface-based analytics software which helps in answer all forms of questions for a mobile application where the user will be able to track very specific data	25,000 free data point limit

8 Benefits, Challenges and Limitations

8.1 Benefits of Mobile Analytics

In order to understand the importance of a process, the understanding of the benefits of the process is important [21]. This section discusses about the potential benefits which can be found by using mobile data analytics.

1. The analysis provides the developers to design a more responsive application.
2. Helps in identifying the prospect of media spend.
3. Helps the developers to make better channel-based marketing and get higher return on investment from the same mobile channel.
4. Stakeholders of the application can make better decisions based on the analytics.
5. Allows the user to study the reach of the paid mobile campaigns.
6. Developers will be able to provide the user with higher customer satisfaction.
7. The analytics of the data helps in understanding the current trend and opportunities in the mobile app market.

8.2 Challenges in Mobile Analytics

The mobile data analytics industry faces a number of challenges which needs to be shared with the readers [22]. The major challenges faced by the mobile analytics industry has been shared below:

1. Unique visitor identification: The data being collected needs to be unique in terms of users who are visiting the app or website. If a single person is coming to the website multiple number of times and doing the same selective number of tasks and swipes then the data being collected would become redundant. New tasks made by unique visitors will helps in understanding where the flaws lie in the current system.

2. App retention analytics: There needs to be a collection of information regarding the number of time a unique user is downloading the same application over and over a certain period of time. This would help in understanding where the user wants to engage mainly on the application and whether the user is only engaging in those same task every time the application is being re-downloaded.
3. Privacy concerns: There needs to be a filter for the data that is being collected from the user and shared with the developers. The user may have some important information that is being saved in the application and does not want it to get sent over to the developers. Similar issues were found with the Google Keyboard application where they were asking for usage statistics to be sent to the developers from time to time in order to provide the users with better next word prediction during their time on the keyboard.

8.3 Limitations in Mobile Analytics

A large quantity of data is being constantly generated by Internet consumers all around the world. However, the data is not being listed in a systematic manner [23]. The area of research can be considered to be predominantly large but the process of analysis in the practical scenario along with the implementation has been found to be difficult in nature. The method of sampling can be considered to be a solution but the data is being generated by both the users as well as the developers. Technical changes are being done on a daily basis in order to achieve the goals set out by the company [23]. The major concern about the amount of data being collected for the process of analysis can be limited by the issues pertaining to data security and the unacceptance of users to share their usage statistics with the developers.

9 Conclusion

On the basis of the study conducted to present this research to the readers, it can be concluded that the use of mobile data analytics is an integral part of sustaining any form of mobile application. The collection of data from the users and then analyzing the data to provide the users with the required update so that the issue they were facing can be resolved are considered to be an important segment in this process. The inclusion of the case studies helps the readers to understand the proper working of mobile analytics that are being used in different mobile development organizations. In addition to these information, the details about the benefits, challenges and limitations of the use of mobile data analytics have been included.

References

1. Thiyagaraj, P.B., Akalya, K., Joicy, I.J.: Mobile data analytics: an overview of tools. IJSRCSAMS **8**(2) (2019)
2. Li, N., Ye, Q.: Mobile data collection and analysis with local differential privacy. In: 2019 20th IEEE International Conference on Mobile Data Management (MDM) (2019)
3. Van Esch, D., et al.: Writing across the world's languages: deep internationalization for Gboard, the Google keyboard, arxiv.org (2019)
4. Guo, L., Sharma, R., Yin, L., Lu, R., Rong, K.: Automated competitor analysis using big data analytics: evidence from the fitness mobile app business. Bus. Process Manag. J. **23**(3), 735–762 (2017)
5. Kannan, S., Rajeswari, S., Suthendran, K., Rajakumar, K.: A smart agricultural model by integrating IoT, mobile and cloud-based big data analytics. In: 2017 International Conference on Intelligent Computing and Control (I2C2)
6. Lv, Z., Song, H., Basanta-Val, P., Steed, A., Jo, M.: Next-generation big data analytics: state of the art, challenges, and future research topics. IEEE Trans. Ind. Inf. **13**(4), 1891–1899 (2017)
7. Parwez, M.S., Rawat, D.B., Garuba, M.: Big data analytics for user-activity analysis and user-anomaly detection in mobile wireless network. IEEE Trans. Ind. Inf. **13**(4), 2058–2065 (2017)
8. Salah, A.A., Pentland, A., Lepri, B., Letouzé, E., De Montjoye, Y.A., Vinck, P.: Guide to mobile data analytics in refugee scenarios. The 'Data for Refugees Challenge' study. Springer, Cham (2019)
9. Shorfuzzaman, M., Hossain, M.S., Nazir, A., Muhammad, G., Alamri, A.: Harnessing the power of big data analytics in the cloud to support learning analytics in mobile learning environment. Comput. Hum. Behav. **92**, 578–588 (2019)
10. Carroll, J.K., Moorhead, A., Bond, R., LeBlanc, W.G., Petrella, R.J., Fiscella, K.: Who uses mobile phone health apps and does use matter? A secondary data analytics approach. J. Med. Internet Res. **19**(4), e5604 (2017)
11. Xu, C., Ren, J., She, L., Zhang, Y., Qin, Z., Ren, K.: EdgeSanitizer: locally differentially private deep inference at the edge for mobile data analytics. IEEE Internet Things J. **6**(3), 5140–5151 (2019)
12. Guo, L., Sharma, R., Yin, L., Lu, R., Rong, K.: Automated competitor analysis using big data analytics: evidence from the fitness mobile app business. Bus. Process Manag. J. (2017)
13. Debolina, G., Singh, J.: A novel approach of software fault prediction using deep learning technique. In: Automated Software Engineering: A Deep Learning-Based Approach, pp. 73–91. Springer, Cham (2020)
14. Minelli, R., Lanza, M.: Software analytics for mobile applications—insights & lessons learned. In: Proceedings of the European Conference on Software Maintenance and Reengineering (CSMR), pp. 144–153 (2013)
15. Abolfazli, S., Lee, M.R.: Mobile data analytics. IT Prof. **19**(3), 14–16 (2017)
16. Sablatura, J., Karabiyik, U.: Pokémon go forensics: an android application analysis. Information (2017)
17. Wagner-Greene, V.R., Wotring, A.J., Castor, T., Kruger, J., Mortemore, S., Dake, J.A.: Pokémon GO: healthy or harmful? Am. J. Public Health (2017)
18. Loveday, P.M., Burgess, J.: Flow and Pokémon GO: the contribution of game level, playing alone, and nostalgia to the flow state. E-J. Soc. Behav. Res. Bus. (2017)
19. Ur Rehman, M.H., Batool, A., Liew, C.S., Teh, Y.W.: Execution models for mobile data analytics. IT Prof. **19**(3), 24–30 (2017)
20. Valizadeh, M.: Using Google Keyboard in L2 writing: impacts on lexical errors reduction. J. Lang. Teach. Learn. **11**(2), 61–80 (2021)

21. Jhalani, R., Sharma, G.: Scope and challenges of mobile analytics in digital learning. J. Manag. Eng. Inf. Technol. **564**(2015), 7–11 (2017)
22. Qi, J., Li, L., Li, Y., Shu, H.: An extension of technology acceptance model: analysis of the adoption of mobile data services in China. Syst. Res. Behav. Sci. **26**(3), 391–407 (2009)
23. Laurila, J.K., et al.: The Mobile Data Challenge: Big Data for Mobile Computing Research (2012)

Method Level Refactoring Prediction by Weighted-SVM Machine Learning Classifier

Rasmita Panigrahi, Sanjay K. Kuanar, and Lov Kumar

1 Introduction

Android apps have firmly recognized themselves as conventional software systems that are widely deployed. Today's scenario dictates that the majority of software corporations will use object-oriented (OO) tools to construct contemporary software systems due to their efficient blueprint characteristics such as reusability (extending the code's use again) and vulnerability reduction, which enables more rapid product development. Code refactoring is the process of modifying the code internal structure without affecting the external output. It reduces the code smell due to code length and code complexity, which occupies more space and time. Different smells are bloater, large class, long method, god class, feature envy, duplicate code, etc. Researchers have used extract class, extract method, push method, inline method, and inline type to compensate for the code smell. For efficient analysis of the developer's emotions, some refactoring activities are organized move method, move class, move attributes, push down method and push down attributes. To exhibit project cloning and commit level extraction, RefTypeExtractor and RefactoringMiner are used. The false-positive rate is evaluated from the real-time data collected from the projects where the individual score is compiled to achieve result from the overall score. The output

R. Panigrahi (✉) · S. K. Kuanar
Department of Computer Science and Engineering, School of Engineering and Technology, GIET University, Gunupur, India
e-mail: rasmita@giet.edu

S. K. Kuanar
e-mail: sanjay.kuanar@giet.edu

L. Kumar
Department of Computer Science and Information System, BITS Pilani Hyderabad Campus, Secunderabad, India
e-mail: lovkumar@hyderabad.bits-pilani.ac.in

J. Singh et al. (eds.), *Mobile Application Development: Practice and Experience*,
Studies in Systems, Decision and Control 452,
https://doi.org/10.1007/978-981-19-6893-8_8

analysis is done to differentiate between commit message and refactoring commit message based on clone GitHub projects. Refactoring has been extensively studied in object-oriented software, but its impact on mobile app quality is still unknown. The authors' [1] work is the first empirical investigation to fill this void to the best of our knowledge. Refactoring operations totaled 42,181 for 300 open-source smartphone apps studied in this large empirical investigation. An inference method established on the Difference-in-Differences (DiD) model was used for causal inference to examine the effect of these refactoring operations on ten standard quality metrics. Refactoring generally leads to better metrics, according to their findings. It is not uncommon for refactoring to have little or no effect on the metrics, but the LCOM metric, in particular, is affected. Refactoring to the context of Android app development can be improved with the help of these findings. When they conducted a large-scale study across eight libraries, they used a tool-based approach (i.e., totaling 183 consecutive versions). According to that article, prior decisions affecting (a) data selection and (b) example labeling are the root causes of bias. The author's [2, 3] Fair-SMOTE algorithm removes biased labels and re-balances internal distributions so that examples in both positive and negative classes are equal. We are all familiar with the concept of bad design and development practices known as "object-oriented code smells" in the software industry. The research community has identified new classes of code smells as mobile-specific because of mobile apps' proliferation. In the context of performance issues or bottlenecks, these code smells serve as indicators. Despite the numerous empirical studies on these new code smells, the spread and evolution of these smells over time remain a mystery despite the numerous empirical studies on these new code smells. Authors [4] examined how Android code smells were introduced, evolved, and eventually removed in a large-scale empirical study. More than 500 smell-removing commits were manually analyzed, and 25 Android developers were interviewed as part of the study. Findings from their research show that the high softness of mobile-specific code smells is not due to a reduction in pressure. Aside from the fact that developers are unaware of smell instances, the researchers discovered that smell removal is an unintended side effect of maintenance activities.

2 Literature Survey

A software artifact's refactoring is an integral part of the maintenance phase of the software life cycle. Automating the software refactoring process at the design and code levels has been proposed by researchers to reduce the time and effort required to accomplish this task. Papers that advocate, present, or implement an automated refactoring process were analyzed in the paper [5].

Refactoring is a vital part of software maintenance and is often used to enforce best practices or deal with design defects. To better understand how developers document their refactoring activities during the software development life cycle, authors [6] have written in their paper. Self-affirmed refactoring is a term used to describe this refactoring in the commit messages. Refactoring-related change messages are extracted from the text and used to identify refactoring patterns. Authors [7] have analyzed 800 open-source projects to see if the previous findings can be applied to a broader range of projects. The most responsible developers for refactoring activities typically hold advanced positions in their development teams, demonstrating their extensive knowledge of the systems they contribute.

The author's [8] research aims to discover how library API clients are affected by refactoring. To better understand the relationship between API breakages and refactorings, we distinguish between library APIs based on their client usage (referred to as client-used APIs). They used a tool-based approach to conduct a large-scale study across eight libraries (i.e., totaling 183 consecutive versions). First, formative research on 611 widely used Android apps was undertaken to map out the landscape of asynchronous Android apps, understand how developers retrofit asynchrony, and learn about the obstacles faced by developers. After following this study, authors [9] developed and tested ASYNCDROID—a refactoring tool that helps Android developers correct incorrectly used async constructs. ASYNCDROID is useful, accurate, and time-saving by the researchers. There were 45 refactoring patches submitted, and developers believe the refactorings are helpful.

The feature extraction method converts raw data into numerical features that can be processed while retaining the original data's information. Therefore, it outperforms applying machine learning directly to raw data. Based on the requirements of small and medium-sized systems, the authors describe their experience refactoring features. A new modularization paradigm called Feature-Oriented Software Development (FOSD) was used to implement the eight refactoring patterns that explain how to extract the elements of features, which were then implemented using FOSD. However, according to the authors, some open issues need to be addressed to automate feature-oriented refactoring [2, 10]. Previously researchers used Wilcoxon rank sum test for relevant feature extraction and selection for refactoring prediction, which gave the best result compared to other techniques. Finally, the extracted features will be fed into a machine learning framework to get the desired result.

Machine learning algorithms are examined by the authors [11] to predict software refactorings. Over two million refactoring's from over 11,149 real-world projects from the Apache and F-Droid ecosystems are used to train six different machine learning algorithms (i.e., Naive Bayes, logistic regression, support vector machine, random forest, decision trees and neural network). Refactorings at the class, method, and variable levels can be predicted with up to 90% accuracy. According to our findings, random Forests are the best models for software refactoring prediction; process and ownership metrics appear to be important in developing better models, and models generalize well across different contexts. The study [extra] uses a weighted support vector machine (weighted-SVM) to solve the outlier sensitivity problem of standard SVMs for two-class data classification. The central concept is to give varying weights to various data points. The WSVM training algorithm learns the decision surface based on the relative importance of the variables. Data points in the training dataset have a lot of weight. Therefore, the weights that are used in WSVM are created. Kernel-based possibility c-means are based on a robust fuzzy clustering technique (KPCM), a partitioning algorithm that produces relatively high values for key data points, yet outliers have low values.

Generally, weighted-SVM is implemented on an imbalanced dataset, so this chapter aims to verify whether SVM with SMOTE is better or weighted-SVM is achieving better results. According to our knowledge, this is the first work of such model construction which describes the feature extraction and selection of relevant features through Wilcoxon rank sum test and refactoring prediction through SMOTE with SVM and weighted-SVM machine learning classifiers with different kernels. Furthermore, boxplot diagrams and descriptive statistics have presented a comparative study regarding the different kernels and classifiers.

RQ1: **Can sampling techniques assist in the improvement of classification models**? We used the oversampling balance technique to compare and evaluate our competitor classification algorithms on the imbalanced datasets.
RQ2: How effective is our machine learning in recommending refactoring based on our baseline sampling approach? Generally, weighted-SVM is applied on the dataset where data is not properly balanced. So we have used SMOTE with SVM to compare the result with weighted-SVM.
RQ3: Can SMOTE with SVM is effectively working as comparison to weighted-SVM? On our dataset SMOTE with SVM is effectively working as compare to weighted-SVM machine learning classifier.
RQ4: **Which Kernel of SVM technique gives better result?** Three (linear, polynomial and RBF) different kernels for each classification technique have been used in this work and classification with RBF kernel gives better results.

3 Background Study

This section describes different research work involved in our chapter, such as refactoring, feature extraction and selection, data sampling and implementation of machine learning framework.

3.1 Refactoring

Refactoring is a well-known and widely used technique for optimizing existing software design, particularly for large-scale and modern software systems that rely on many third-party libraries. Fowler recommends refactoring to increase the readability and reuse of software while simultaneously increasing the speed with which developers can write and maintain their code base. In addition, refactoring addresses the issue of architectural degradation in object-oriented software projects by improving the project's internal structure without modifying the behavior. On the other hand, identifying refactoring opportunities is a complex problem for developers and researchers alike. However, recent research has demonstrated that machine learning algorithms can solve this problem.

3.2 *Feature Extraction and Selection*

The feature extraction method transforms raw data into numerical features that can be processed while preserving the integrity of the original data. After preprocessing and cleaning the data, the Wilcoxon rank test has been applied to specify the predictability of refactoring. In total 67 metrics have been computed for each project. As all the metrics are unnecessary, irrelevant metrics should be removed except the important metrics required for future prediction. The SMOTE data sampling technique has been applied to balance our dataset, which has been considered from the tera-PROMISE repository. In hypothesis testing modules, we can learn how to test for mean equality between two different samples. When comparing two independent samples whose results are not normally distributed, a non-parametric test is appropriate. Numerous researchers have observed that the t-test is less suitable for unbalanced data than the Wilcoxon rank sum test. Based on the previous researchers point of view, the Wilcoxon rank sum test has been selected as a feature selection technique in this work. Many researchers have already observed that a well-chosen set of metrics produces superior results.

3.3 *Classification Techniques: SMOTE with SVM and Weighted-SVM*

Classification and regression problems can be solved using support vector machines (SVMs), one of the most common supervised learning methods. The SVM algorithm's objective is to find the optimal line or decision boundary that partitions n-dimensional space into classes, allowing us to easily classify new data points in the future. This optimal decision boundary is referred to as a hyperplane. SVM selects the hyperplane's extreme points/vectors. Support vectors to refer to these extreme circumstances, and the method is referred to as a support vector machine. Although the support vector machine technique effectively classifies balanced datasets, it struggles with imbalanced datasets. A hyperplane decision boundary is found via the SVM method. A margin allows some points to be misclassified, softening the divide. This margin favors the majority class by default with skewed class distributions, but it can be changed to benefit all classes equally. SVM is among the most robust and accurate classification algorithms. The kernel trick allows SVMs to perform nonlinear classification efficiently by implicitly mapping inputs into high-dimensional feature space. This model is called a support vector machine, or SVM, used to solve classification and regression problems. Linear and nonlinear problems can be solved using this tool. SVM is a simple concept: As a result of this algorithm, the data is divided into distinct classes.

SVM technique utilizes the idea of deriving an exclusive splitting hyperplane (optimal hyperplane) which will maximize the margin between the refactoring and non-refactoring classes. Training data points are expressed as l

$$\{(x_i, y_{i)}\}\frac{l}{i=1}, \quad x_i|\epsilon R^N, y_i\epsilon\{-1, 1\} \tag{1}$$

The following optimization problem can be solved by support vector technique:

$$\text{Minimize } \phi(w) = \frac{1}{2}w^T w + C\sum_{i=1}^{l}\epsilon_i \tag{2}$$

focus to

$$y_i(\langle w, \phi(x_i)\rangle) + b \geq |1 - \epsilon_{i,},$$
$$\epsilon_i \geq 0, \quad i = 1, \ldots, l \tag{3}$$

Weight vector w_0 can be maximized by using this formula:

$$w_0 = \sum_{i=1}^{l}\alpha_i y_i \phi(x_i) = \sum_{i=1}^{l_s}\alpha_i y_i \phi(x_i), \tag{4}$$

Decision functions can be derived from the optimal pair (w0-b0) once the pair has been identified as:

$$f(x) = \text{sign}(\langle w_0, \phi(x)\rangle b_0) = \text{sign}\left(\sum_{i=1}^{l_s} a_i y_i K(x, x_i) + b_0\right) \tag{5}$$

According to the previous equation of $f(x)$, only the points with $I > 0$ can determine the decision boundary. However, many practical engineering applications have noises or outliers in the training data, serving as support vectors for SVM training. In this case, the optimal decision boundary has strayed. The standard SVM algorithm has an outlier sensitivity problem. The outlier sensitivity of SVM is illustrated in Fig. 1. If the training data contains no outliers, an SVM can find an optimal hyperplane (solid line) with a maximum margin to separate the two classes. However, the decision hyperplane (dotted line) has deviated significantly from the optimal if two outliers (numbered 1 and 2) are present.

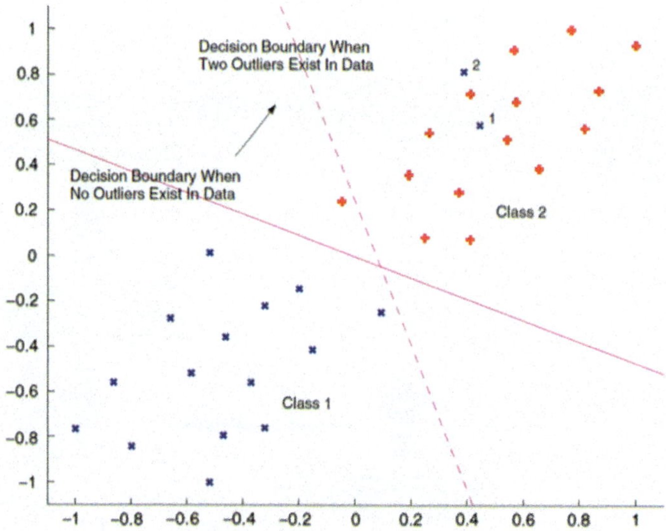

Fig. 1 Extremely sensitive to outliers in SVM

4 Proposed Methodology

Sometimes, Android app development takes the help of programming for solving the runtime issues, so refactoring can be implemented. Three phases have been followed for refactoring prediction at the method level in our proposed methodology. Five publicly available Java projects (Antlr4, oryx, titan, mct, Junit) from the tera-PROMISE repository, validated by the author [12], have been considered to carry out our target. In the first phase, source meter stool is used to find the total 65 metrics from each project. Then the Wilcoxon rank sum test is then applied to identify the relevant source code metrics required for refactoring the prediction machine learning framework.

The computed 67 source code metrics will have the capability to decide whether they can help to refactor the method or not. Second phase is the data sampling to address the class imbalance issue. The SMOTE technique is the best suitable for balancing the dataset, proved by different researchers [13]. After balancing, the sampled dataset has to be considered for feature selection as next process. The relevant features selected for refactoring prediction will be considered input for the machine learning framework. Weighted-SVM is utilized to predict the need for method refactoring as the 3rd phase. Finally, The efficiency of the proposed model has been measured in terms of different types of performance metrics (i.e., AUC and accuracy) and shown through boxplot diagram. All three phases are shown in Fig. 2.

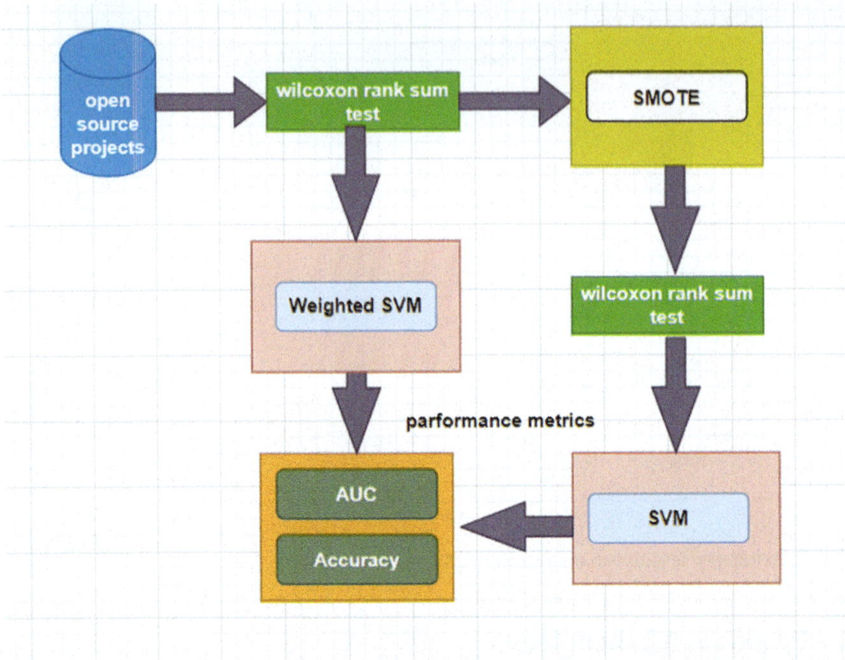

Fig. 2 Proposed methodology for refactoring prediction using weighted-SVM as well as SMOTE with SVM

5 Comparative Analysis

This section shows comparison analysis through boxplot as well as descriptive statistics. We have considered the original data and sampled data for refactoring prediction with all metrics and significant metrics. The overall performance of SMOTE with SVM and weighted-SVM (WSVM) is shown on Table 1.

5.1 Classifier's Significance

This chapter validates that weighted-SVM with imbalanced data gives the lower result as comparison to SVM with SMOTE technique.

In Table 2, we can visualize that mean AUC value of SMOTE-SVM is 0.90 and mean AUC value of weighted-SVM is 0.42. The mean accuracy of weighted-SVM is 89% and SMOTE-SVM is 82%.

This study uses a rank sum test to determine the impact of SMOTE techniques on refactoring prediction model performance over original data. By considering the null hypothesis, "In terms of software refactoring model prediction, the SMOTE

Table 1 AUC and accuracy performance of weighted-SVM and SMOTE with SVM

				AUC			Accuracy		
				LIN	POLY	RBF	LIN	POLY	RBF
antlr4	ORG	AM	WSVM	0.54	0.33	0.52	87.15	96.98	92.11
Junit	ORG	AM	WSVM	0.6	0.26	0.49	73.78	83.71	85.12
Mct	ORG	AM	WSVM	0.27	0.29	0.15	87.43	70.78	91.76
Oryx	ORG	AM	WSVM	0.64	0.26	0.75	85.67	96.98	92.94
Titan	ORG	AM	WSVM	0.44	0.45	0.45	88.98	97.19	89.38
antlr4	ORG	SG	WSVM	0.5	0.26	0.5	87.63	96.41	91.13
Junit	ORG	SG	WSVM	0.38	0.25	0.32	72.85	89.89	82.56
Mct	ORG	SG	WSVM	0.27	0.25	0.28	87.05	96.75	92.99
Oryx	ORG	SG	WSVM	0.61	0.34	0.68	83.08	92.86	92.99
Titan	ORG	SG	WSVM	0.48	0.51	0.55	88.48	98.39	91.43
antlr4	SMOTE	AM	SVM	1	1	1	100	100	100
Junit	SMOTE	AM	SVM	0.89	0.94	0.99	80.6	84.75	91.48
Mct	SMOTE	AM	SVM	0.87	0.92	0.95	81.1	73.48	89.15
Oryx	SMOTE	AM	SVM	0.95	0.96	0.98	88.58	83.24	93.76
Titan	SMOTE	AM	SVM	0.83	0.8	0.89	69.81	67.86	79.64
antlr4	SMOTE	SG	SVM	0.88	0.92	0.96	80.96	80.86	90.22
Junit	SMOTE	SG	SVM	0.81	0.87	0.96	69.04	73.39	85.78
Mct	SMOTE	SG	SVM	0.79	0.88	0.93	74.59	73.95	84.28
Oryx	SMOTE	SG	SVM	0.91	0.94	0.96	84.51	81.4	88.96
Titan	SMOTE	SG	SVM	0.74	0.75	0.84	67.25	68.52	72.88

Table 2 Descriptive statistics of all the classifiers and kernels

		Min	Max	Mean	Q1	Q3
AUC	LIN	0.27	1	0.67	0.49	0.88
	POLY	0.25	1	0.61	0.28	0.92
	RBF	0.15	1	0.71	0.49	0.96
Accuracy	LIN	67.25	100	81.93	74.18	87.53
	POLY	67.86	100	85.37	73.71	96.87
	RBF	72.88	100	88.93	85.45	92.53
AUC	WSVM	0.15	0.75	0.42	0.27	0.52
	SMOTE-SVM	0.74	1	0.9	0.87	0.96
Accuracy	WSVM	70.78	98.39	88.81	85.67	92.99
	SMOTE-SVM	67.25	100	82	73.48	88.96

Table 3 Classifier's significance

	WSVM	SMOTE-SVM	Mean-Rank
WSVM	1	3.32E−11	2
SMOTE-SVM	3.32E−11	1	1

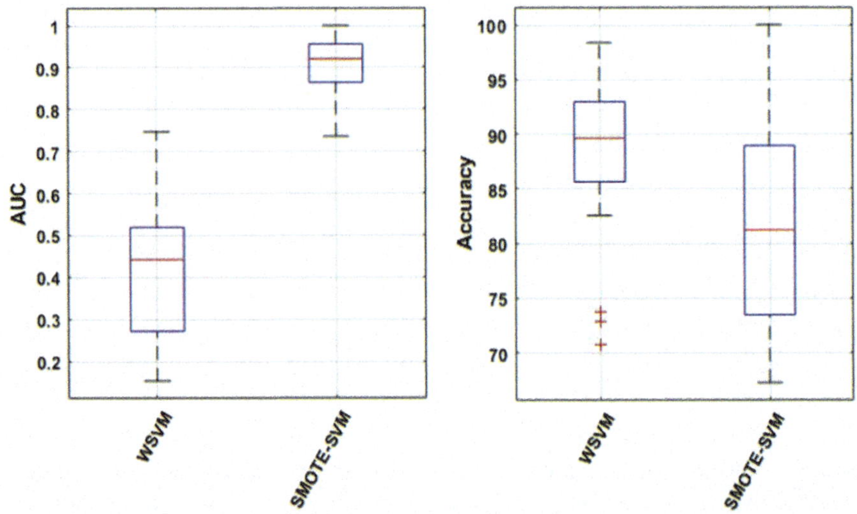

Fig. 3 Boxplot illustrating outliers, the percentile value, the median, the interquartile range and the degree of dispersion for classifier's accuracy values

sampled data with SVM are no better than the original data with weighted-SVM." The standardized significance level of 0.05 was used to ensure the hypothesis was true. If the p-value ≤ 0.05, the null hypothesis is rejected and vice versa. From Table 3, we can observe that all the generated values are more than 0.05 and all the generated values are significantly same, which means the NULL hypothesis is rejected. To make differentiate between two classification techniques, a Friedman test has been computed for finding mean-rank. Table 3 shows that SMOTE with SVM secures 1 mean-rank, whereas the mean-rank of weighted-SVM is 2, which proves SMOTE with SVM has better results than weighted-SVM. Its comparison has been shown in terms of boxplot diagram in Fig. 3.

5.2 *Kernel's Significance*

This chapter has implemented three kernels (linear, polynomial, RBF) for each classification technique. Descriptive statistics of kernels shows man AUC value for RBF

Table 4 Kernel's significance

	LIN	POLY	RBF	Mean-Rank
LIN	1	0.57	0.36	2.35
POLY	0.57	1	0.19	2.325
RBF	0.36	0.19	1	1.325

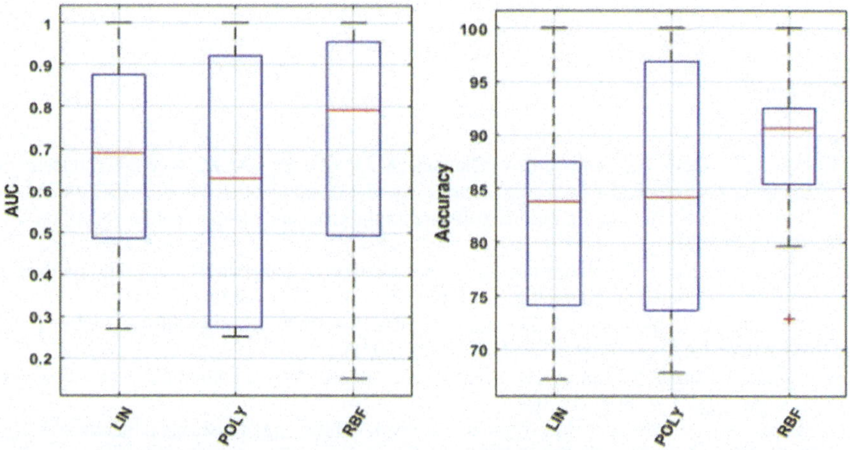

Fig. 4 Boxplot illustrating outliers, the percentile value, the median, the interquartile range and the degree of dispersion for Kernel's accuracy values

is 0.71 and mean accuracy is 88.93, which is more than the other Kernel's performance shown in Table 2. The Kernel's significant test has been conducted by taking a NULL hypothesis. "There is no difference between all the kernels of classification techniques." In Table 4, we can observe the maximum values are more than 0.05 and significantly same. So the hypothesis is rejected. That is why mean-rank has been computed by Friedman test of all the kernels. Table 4 concludes that RBF kernel has 1.325 mean-rank, which is very less. So RBF kernel attains better result as compare to other. All the kernels boxplot diagrammatic comparative study has been shown in Fig. 4.

Table 4 describes the mean-rank of RBF kernel is very less, i.e., 1.325, and it signifies that RBF kernel achieves good results compared to other kernels.

6 Conclusion

In Android apps, there is a chance of code smells for which refactoring can be applied to improve its quality. A SMOTE with SVM for refactoring prediction is proposed as a solution to the problem of refactoring prediction in comparison with weighted-SVM.

Previously researchers have proved that weighted-SVM is meant for only imbalanced dataset. Still, it hs been proved in our chapter that SMOTE with SVM (balanced data) achieves better accuracy than weighted-SVM with imbalanced data. SVM with different kernels can be implemented for the refactoring prediction purpose, but RBF kernel gives the best result out of all the kernels.

References

1. Hamdi, O., Ouni, A., AlOmar, E.A., Cinnéide, M.O., Mkaouer, M.W.: An empirical study on the impact of refactoring on quality metrics in android applications. In: 2021 IEEE/ACM 8th International Conference on Mobile Software Engineering and Systems (MobileSoft), pp. 28–39. IEEE (2021)
2. Singh, J., Khilar, P.M., Mohapatra, D.P.: Code refactoring using slice-based cohesion metrics and aspect-oriented programming. Int. J. Bus. Inf. Syst. **27**(1), 45–68 (2018)
3. Chakraborty, J., Majumder, S., Menzies, T.: Bias in machine learning software: why? How? What to do? arXiv preprint arXiv:2105.12195 (2021)
4. Habchi, S., Moha, N., Rouvoy, R.: Android code smells: from an introduction to refactoring. J. Syst. Softw. **177**, 110964 (2021)
5. Baqais, A.A.B., Alshayeb, M.: Automatic software refactoring: a systematic literature review. Softw. Qual. J. **28**(2), 459–502 (2020)
6. AlOmar, E., Mkaouer, M. W., Ouni, A.: Can refactoring be self-affirmed? An exploratory study on how developers document their refactoring activities in commit messages. In: 2019 IEEE/ACM 3rd International Workshop on Refactoring (IWoR), pp. 51–58. IEEE (2019)
7. Alomar, E.A., Peruma, A., Mkaouer, M.W., Newman, C.D., Ouni, A.: Behind the scenes: on the relationship between developer experience and refactoring. J. Softw.: Evol. Process e2395 (2021)
8. Gaikovina Kula, R., Ouni, A., German, D.M., Inoue, K.: An empirical study on the impact of refactoring activities on evolving client-used APIs. arXiv e-prints, arXiv:1709 (2017)
9. Lin, Y., Okur, S., Dig, D.: Study and refactoring of android asynchronous programming (T). In: 2015 30th IEEE/ACM International Conference on Automated Software Engineering (ASE), pp. 224–235. IEEE (2015)
10. Lopez-Herrejon, R.E., Montalvillo-Mendizabal, L., Egyed, A.: From requirements to features: an exploratory study of feature-oriented refactoring. In: 2011 15th International Software Product Line Conference, pp. 181–190. IEEE (2011)
11. Aniche, M., Maziero, E., Durelli, R., Durelli, V.: The effectiveness of supervised machine learning algorithms in predicting software refactoring. IEEE Trans. Softw. Eng. (2020)
12. Kádár, I., Hegedus, P., Ferenc, R., Gyimóthy, T.: A code refactoring dataset and its assessment regarding software maintainability. In: 2016 IEEE 23rd International conference on software analysis, Evolution, and Reengineering (SANER), vol. 1, pp. 599–603. IEEE (2016)
13. Panigrahi, R., Kumar, L., Kuanar, S.K.: An empirical study to investigate different SMOTE data sampling techniques for improving software refactoring prediction. In: International Conference on Neural Information Processing, pp. 23–31. Springer, Cham (2020)

OCEANDROID

Unnati Shah, Vishruti Desai, Hardi Vyas, Susmita Sonawane, Nirali Modi, and Prem Desai

1 Introduction

Fishing applications can be used for a variety of purposes in research, including regular monitoring and increasing the efficiency, spatial and temporal scope and resolution of traditional survey methods. India's fishing sector makes a significant contribution to the country's economy. It gives millions of individuals' crucial foreign exchange and jobs. At the same time, it is a source of income for a huge portion of the country's economically disadvantaged population. Capture fisheries and aquaculture provide a living for more than 7 million people in the country. Indian fisheries make up a significant portion of global fisheries. India is the world's fourth largest fish producer and the second greatest producer of inland fish. India's contribution to global fish production has risen from 3.2% in 1981 to 4.5% today. The fishing industry plays an essential role in the country's socioeconomic development. However, fishermen face difficulties [1] such as accidentally trespassing past country borders because they do not know where the boundary is, causing citizenship issues and maybe being declared criminals and being unable to identify natural resources.

As a result, OCEANDROID, our suggested android application, addresses the issue of fisherman safety. Furthermore, the suggested application assists fishermen in executing effective fishing operations, such as determining the fish density and route notification. The fishers must set the source and destination to obtain the route to make the program simple to use. We employ GPS technology to accomplish

U. Shah (✉)
Computer Science, Utica University, Utica, USA
e-mail: unshah@utica.edu

V. Desai · H. Vyas · S. Sonawane · N. Modi · P. Desai
Surat, India

J. Singh et al. (eds.), *Mobile Application Development: Practice and Experience*,
Studies in Systems, Decision and Control 452,
https://doi.org/10.1007/978-981-19-6893-8_9

Fig. 1 Objectives of the application

this goal. When fishermen, on the other hand, choose a different route or trespass, the border alarm will sound and they will receive a notification. The software will feature a single push-button for fishermen to press if the boat is hijacked, and the admin and company will be notified. Fish density in a certain area will be assessed and reported, allowing them to fish more easily. There will also be a weather chart and a route tracing tool (reverse navigation). As shown in Fig. 1, the objectives of the developed application are:

1. The safety of fishermen is ensured by issuing a warning if they infringe on the LOC, as well as sending a message to the admin if something threatening occurs, such as a terrorist hijacking of a ship.
2. Fishermen can readily determine the density (number) of fish to target only those areas with the highest concentrations of fish, as well as useful weather charts, route planning and reverse navigation.
3. The in-app camera is used to send information to the admin and other fishermen. Those fishermen who are visiting that particular route can use the in-app camera to take images of any contaminated water they find along the route, as well as the type of fish.

By detecting fish, this value depicts the amount of fish present (Fig. 2) so that fishermen can estimate the number of fish and go easy fishing in areas where there are more fish.

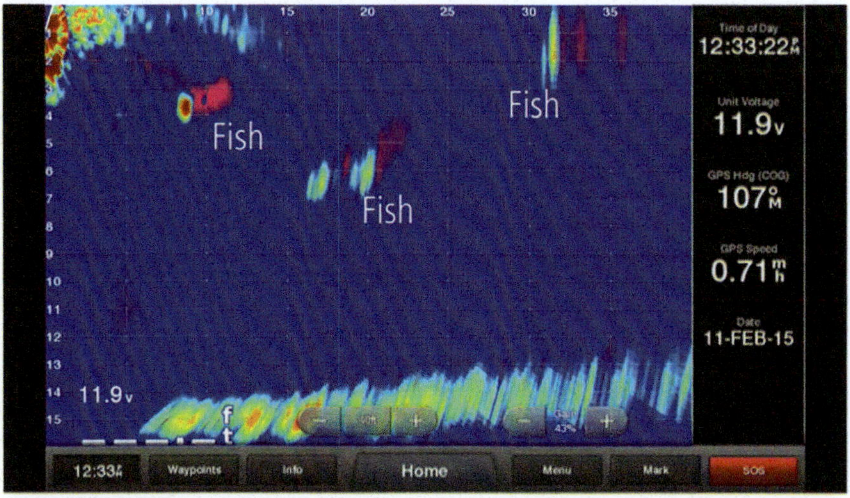

Fig. 2 Display of fishes

2 Literature Review

Table 1 shows the various strategies we discovered in the literature.

The bait trap [1] has a square shape to it. More than one trap is set in line to catch fish in places like Australia and New Zealand. Beam Trawl, on the other hand, is a widely utilized fishing technique [2]. Steel pipe (beam) holding the net open is put on one side of the boat net. However, it has a downside in that it consumes a lot of oil. The other approach is the Cast Net (throw net technique) [3], a fishing throw net. It is a circular net with little weights strewn around the outside. The net is cast by hand in such a way that it floats and stretches out on the water. Furthermore, the fly shooter [4] technique, in which long thick lines are used to place nets, and fish are caught as the ship moves forward. The narrow mesh employed in this approach makes it difficult to catch larger fish, which is a disadvantage.

3 OCEANDROID Application

The application is divided into four modules, viz. (i) route module, (ii) density of fish module, (iii) camera module and (iv) weather forecast module. Figure 3 shows a use-case diagram of the proposed application OCEANDROID, which shows the high-level functionality. The activity of the fishermen and the administrative staff is depicted in this diagram. Login, Validate and Verify, Fishing, Set Route, Weather Forecast, Fish Density, Notify and Alert are some of the use cases.

Table 1 Literature review table

Cite	Techniques	Features	purpose	Limitations/dis
[2]	Bow-fishing, gigging, speargun	• Target a specific fish, hence, not harmful for other fishes • Environment friendly	• Specialized archery equipment to shoot and retrieve fish • Speargun technique is useful for underwater fishing	• Need more practice for the perfect aim • Difficult at night time • Need clear water • Need to be nearer to the target • Need to nail your arrow • Spearguns are banned in some areas
[3]	Bait fishing, Cast Net	• Bait fishing attracts fish from far and wide. Hence, one can easily set our rod up and just wait for something to bite • Good for beginners • Cast Net kit is inexpensive and can catch at least any fish in every cast	• Bait fishing: natural baits are living critters that are used to attract fish to hook • Cast Net: circular net in which one can throw a net in shallow water to catch fish by falling and closing on them	• Bait needs special storage considerations like refrigeration or circulating water • Bait attracts many varieties of fish, even small ones and non-selective fishes • Cast Net fishing needs practice for throwing a net • Cast Net catches non-selective fishes
[4]	Fly shooting, Longline drifting	• Catch up to 10 times more than inshore fishing vessels	• Catch big fish like sharks	• Longline fishing techniques also hook many other non-selective mammals
[5]	Combination of all	• Shows the maximum density of fish under the water • Provides a clear view of underwater because it eliminates the water surface glare	• Detect fishes under the water	• Costly for installation

Route Module

This module is concerned with the safety of fishermen, and thus, before going on a fishing expedition, fishermen must first plan a route in our application. They will be warned that they have wandered more than 5 km from the path they have created and should return to it because reverse navigation is accessible. The interface for the same is shown in Fig. 4.

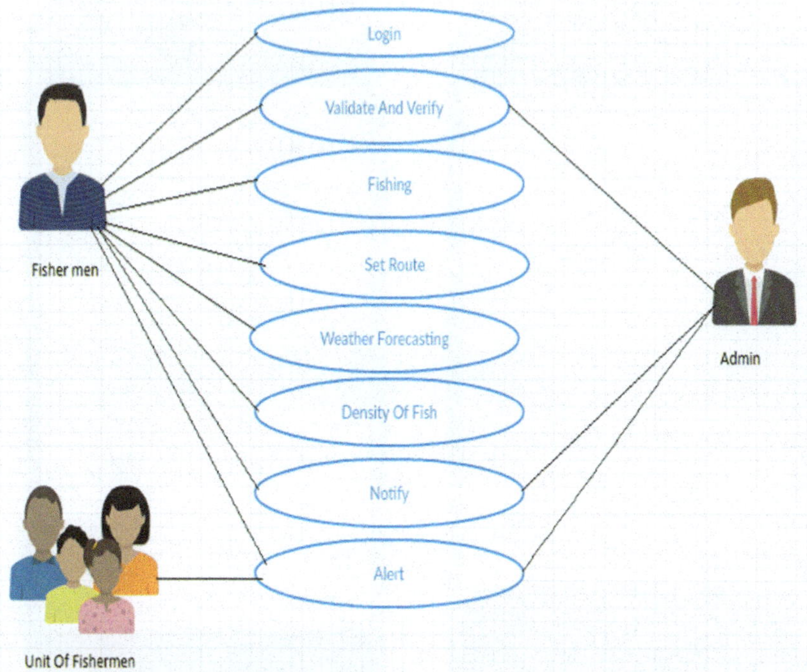

Fig. 3 Use-case diagram for the OCEANDROID

The Density of Fish Module

This module incorporates an IR sensor (used to identify living objects), Bluetooth, capacitors, registers and a converter to make fishing easier and more efficient for fishermen. The reading supplied via Bluetooth to our app will be stable at first, but as fish pass by, the reading will fluctuate often, allowing us to send out notifications that fish are passing through (Fig. 5).

Camera Module

This is an informational module, as seen in Fig. 6. The in-app camera is used to relay data to the admin and other anglers who are passing via that route. Anglers that visit that route can use the in-app camera to photograph any contaminated water they come across as well as the types of fish they encounter.

Weather Forecast Module

This module includes a tidal chart as well as notifications for natural disasters (Fig. 7).

The major goal of our app is to ensure fishermen's citizenship as well as easy fishing as shown in Fig. 8.

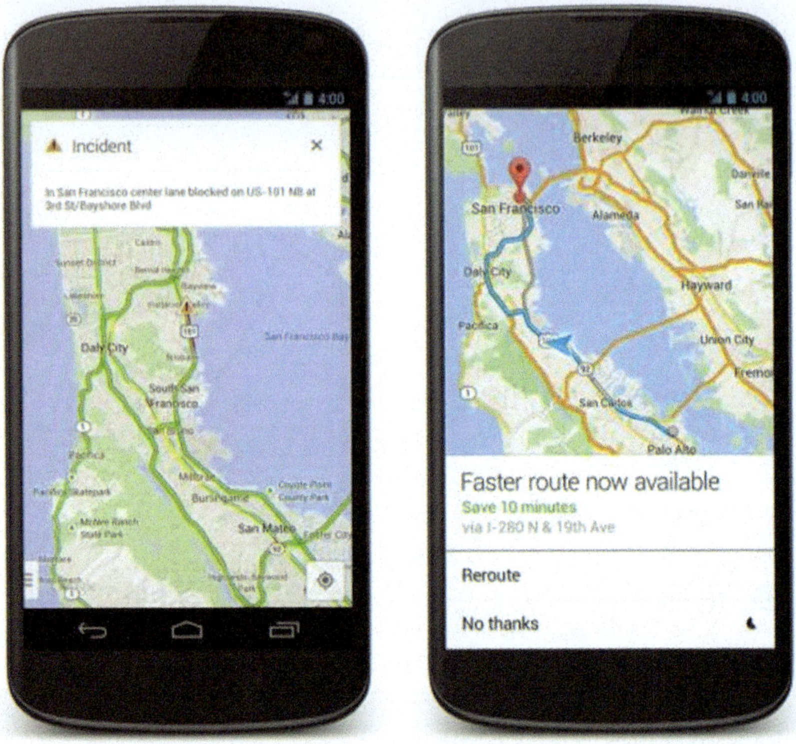

Fig. 4 Route module

It revolves not only around the fisherman's safety from natural disasters but also about violating the border Line Of Control (LOC) and some productivity features that will help them grow their business.

3.1 *Materials/Tools Required*

We used android, PHP, Java, ASP.net and MySQL to build the application. The IR sensor is used. The **IR** is emitted by a **IR** light emitting diode (LED) and received by photodiode, phototransistor or photoelectric cells. During the process of detection, the radiation is altered, between the process of emission and receiving by the object of interest. Figure 9 shows the circuit diagram for the density of the fish module.

The hardware will calculate the density of fish, alert if nearer to the maritime boundary, weather forecast including tides and ebbs chart, suggestion for already visited road and route will be shown on the map. In Table 2, we provide the details of the hardware and software used for the application development.

Fig. 5 Density of fish module

3.2 Results

The results of the developed OCEANDROID application are shown in Fig. 10. The partial source code is available at https://github.com/UnnatiS/OCIENDROID.

3.3 Implications

The application that is being proposed OCEANDROID can determine the density of fish, present a route map, notify you on your way to the route or of the route and contact you in an emergency. By increasing the scope and frequency of sustainable fisheries management, the proposed application can be used to gather data about the spread of aquatic invasive species. Furthermore, the OCEANDROID enables quick reporting and identification of fish movement data, which aids in the discovery

Fig. 6 Camera module

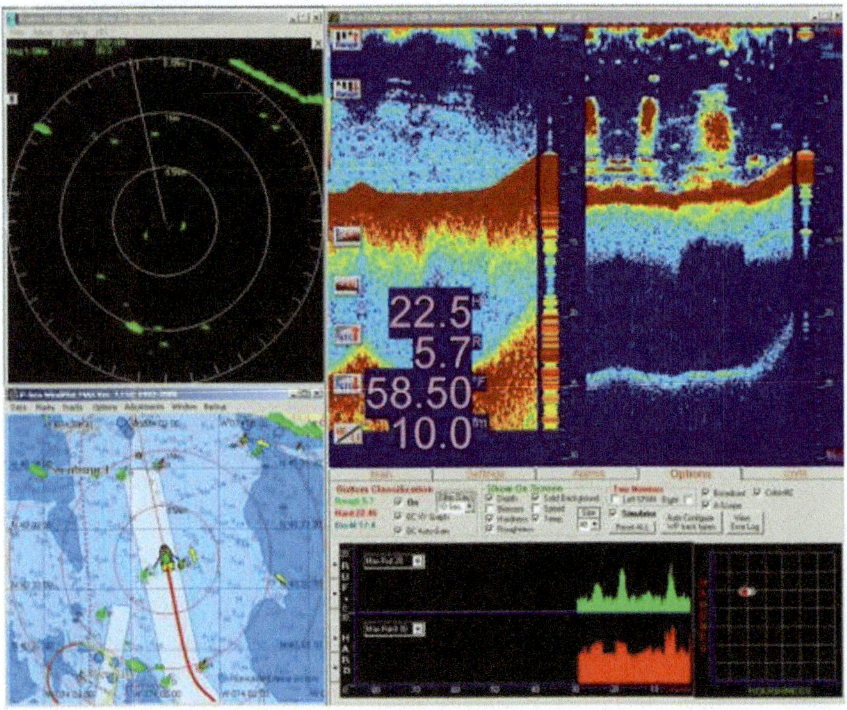

Fig. 7 Weather forecast module

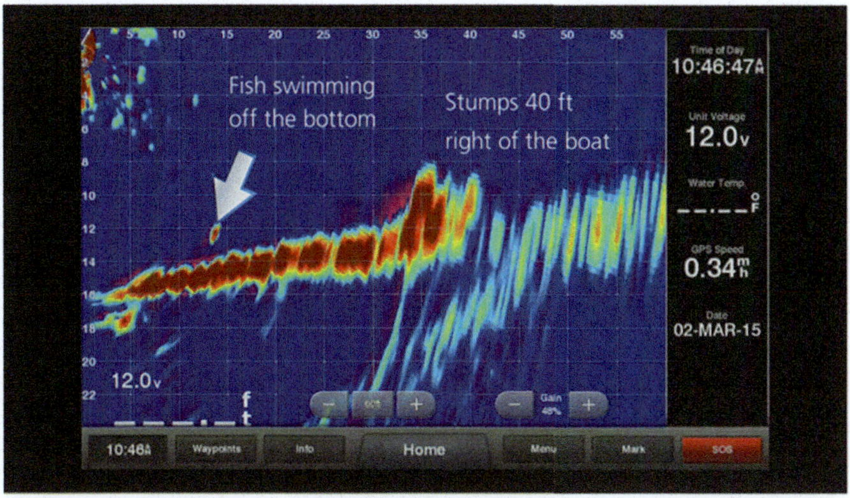

Fig. 8 Fish detection

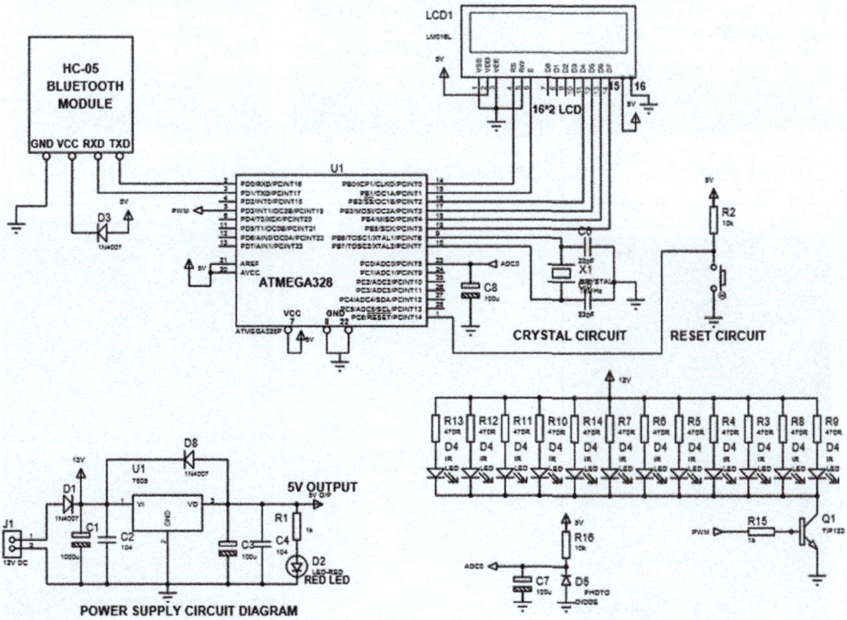

Fig. 9 Circuit diagram of density of fish module

Table 2 Hardware and software used for the application

Hardware components	Software tools
1. Crystal circuit	Android Studio
2. LCD	Visual Studio 2013
3. Bluetooth module	SQL Server Manager
4. Convertor	Internet Information Services (IIS) Manager

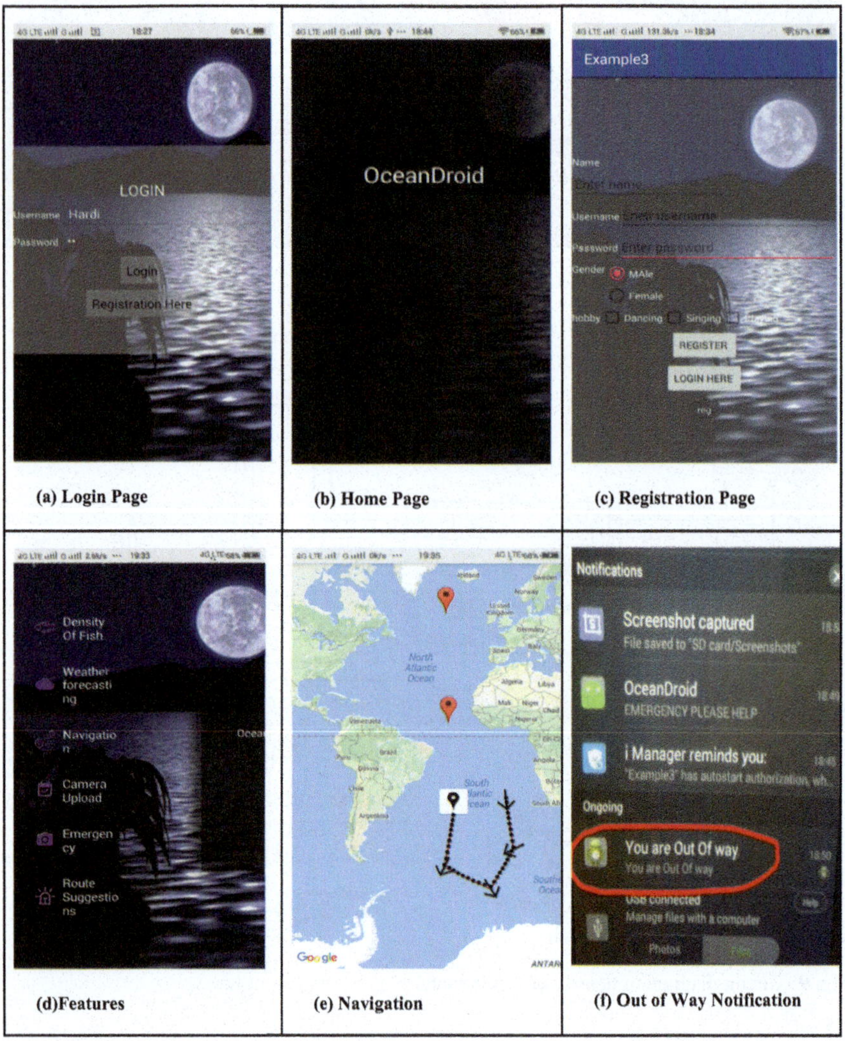

(a) Login Page (b) Home Page (c) Registration Page

(d)Features (e) Navigation (f) Out of Way Notification

Fig. 10 OCENDROID: running application

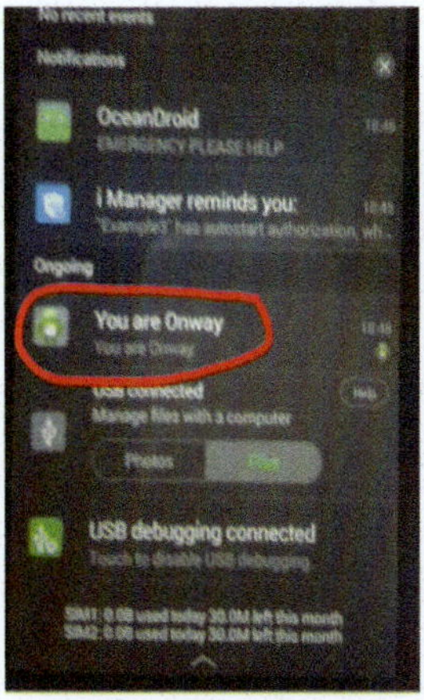

(g) On Way Notification

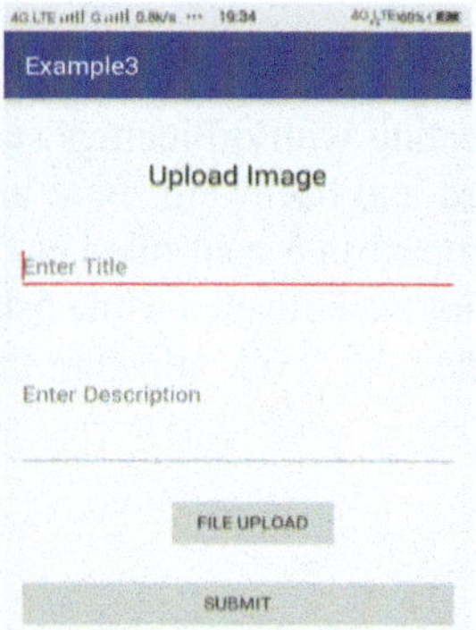

(h) Camera upload

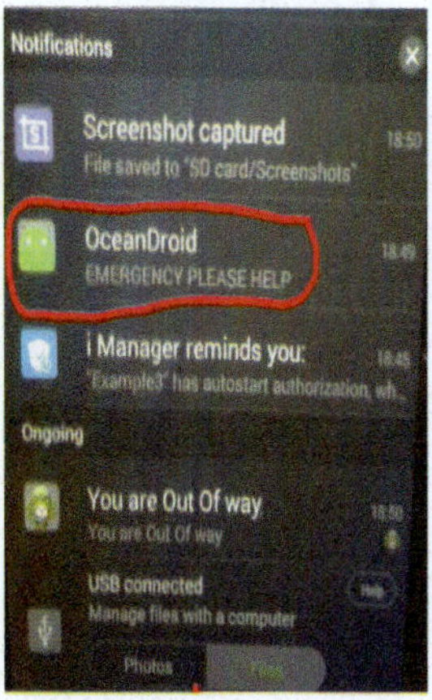

(i) Emergency

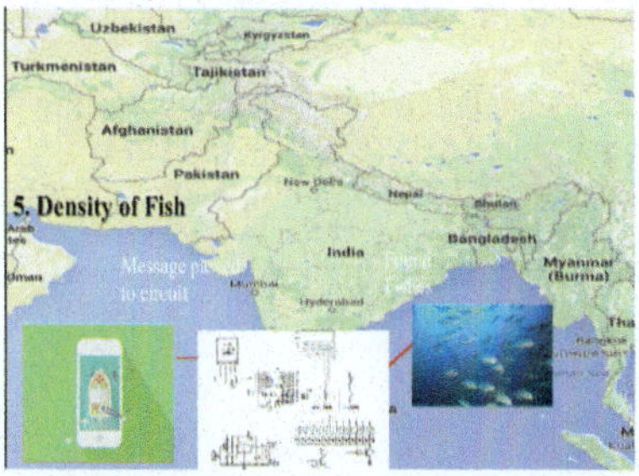

(j) Density of Fish

Fig. 10 (continued)

of paths. Our findings show that the data from our app can meet the demand for quick, low-cost and high-resolution information on fish density, any aquatic invasive species, not just fishes.

In general, the data from our high-resolution, real-time OCEANDROID app opens up intriguing new possibilities for fishing based on weather casting. Furthermore, we can use the fish density data for analysis purposes, such as determining what the fish density was in the past and predicting future density based on that. As a consequence, such statistics can serve as a starting point for determining the likelihood of catching fish. Using the OCEANDROID application instead of traditional methods, which are costly and time-consuming, may help fishermen and fishing organizations to execute their jobs more safely and efficiently. Because our OCENADROD application is evaluated in a controlled experimental environment, its real-time applicability must be confirmed. The experimental results, on the other hand, have a favorable impact. To achieve these goals, we urge that fishing authorities work in the future to create and test the OCEANDROID application.

4 Conclusion

We devised a strategy for dealing with difficulties such as fishermen's safety, fish density detection and forward and backward route information. Our studies show that our app has a positive impact in a controlled environment. We believe that by tracing the whole path the fisherman traveled while fishing and returning navigation, we can elevate security to the next level, allowing the user to enjoy good fishing while feeling secure.

References

1. Zhou, S., Smith, A.D., Knudsen, E.E.: Ending overfishing while catching more fish. Fish Fish. **16**(4), 716–722 (2015)
2. Bear, F.: Underwater Bowhunting. The Archer's Bible (revised ed.), pp. 123–129. Doubleday, New York (1980). ISBN: 0-385-15155-1
3. Mannaa, R.K., Dasb, A.K., Krishna, R.D., Karthikeyanc, M., Singh, D.N.: Fishing crafts and gear in river Krishna (2011)
4. Grieve, C., Brady, D.C., Polet, H.: Best practices for managing, measuring and mitigating the benthic impacts of fishing—Part 1. Mar. Stewardship Council Sci. Ser. **2**, 18–88 (2014)
5. Meenakumari, B. (ed.): Handbook of Fishing Technology. Central Institute of Fisheries Technology (2009)

Innovation Propensity of Firms and the Interplay of Institutional Ecosystem—A Longitudinal Study from G-20 Middle-Income Countries

Debasish Das and Subhasree Mukherjee

1 Introduction

In the domain of innovation strategy, considerable research work has been done to identify various determinants of innovation as a measure of innovation propensity of firms. Scholars [1] have consistently used patent data to analyze various aspects of innovation strategy which includes innovation propensity as well as innovation outcome. Earlier research [2] has also studied the relationship of macro-economic factors, e.g., GDP on innovation strategies of firms. Some other studies have also discussed the impact of innovation strategies on patenting behavior of firms [3]. In literature, we find many such studies, which have delved into various factors impacting innovation strategy. However, most of these studies focus their attention on conventional macro-economic factors, e.g., GDP or common institutional policies, e.g., subsidies, intellectual property awareness, etc. as discussed earlier. Another fact to note is that most of the earlier research were conducted in major economic countries in the European Union [4], OECD [5], Japan, South Korea [6], etc., very little research has been done for the middle-income countries of the world. Although similar scholarly research [7] has also been done for member countries in powerful geo-political economic block, e.g., G20, however, it is negligible to almost no research has been done on the impact of institutional ecosystem factors on innovation activities, specifically, in middle-income economies within the G20

Present Address:
D. Das (✉) · S. Mukherjee
Department of Strategic Management, Indian Institute of Management, Ranchi, India
e-mail: debasish.das20eph@iimranchi.ac.in

S. Mukherjee
e-mail: subhasree.mukherjee@iimranchi.ac.in

J. Singh et al. (eds.), *Mobile Application Development: Practice and Experience*,
Studies in Systems, Decision and Control 452,
https://doi.org/10.1007/978-981-19-6893-8_10

block of nations. Hence, we identify two critical research gaps which provides us an opportunity to analyze the impact of institutional ecosystem on innovation activities within the geo-politically important middle-income member countries of the G20 block. Since firms in each of these middle-income economies have to navigate various institutional factors as a whole, we believe it provide us an opportunity to adopt a more holistic approach to understand impact of institutional ecosystem as a whole and thereby address this research gap in literature. We also focus our attention on the all the nine middle-income economies in the G20 block of nations, which until now have not been studied, thereby, filling the other critical research gap in literature. Argentina, Brazil, China, India, Indonesia, Mexico, Russia, South Africa and Turkey are the nine member countries of the G20 block of nations, and these countries are also categorized as middle-income economies as per data from the [8]. Hence, we focus our research study on all the above nine countries, referred to as G20-MIC countries in the subsequent pages.

As these G20-MIC countries drive global economic growth, naturally there is a significant need for innovation-driven activities in these countries. Many G20-MIC countries are taking multiple steps to foster innovation through changes in regulations, awareness campaigns and other incentives. As one such example is the recently notified government guidelines in India [9] related to examination of Computer Related Inventions (CRIs). These guidelines are expected to reduce patent filing ambiguities for CRIs. Similarly, other countries are also trying to enhance their ecosystem for supporting innovation activities. However, not much research seems to have been done to understand the impact of multiple institutional factors considered as a whole, on the innovation activities in these countries. Therefore, it is important to investigate the impact of existing as well as changes to the institutional ecosystem especially in G20-MIC countries. In this paper we follow methods used by earlier scholars [1] who have used patent filing data as a proxy to measure innovation activity or innovation propensity. And we build hypotheses to investigate the impact of institutional ecosystems factors as a whole to understand their interplay any in patent filing activities in these G20-MIC emerging economies.

The uniqueness of our approach is that we not only focus on conventional macro-economic factors like GDP, or only a few institutional factors, but rather we try to investigate causality between patent filing as a proxy to measure innovation activity and the overall institutional ecosystem which is an aggregation of institutional ecosystem factors consisting of various domestic rights, laws, policies and awareness levels in each of these emerging middle-income economies. Another unique aspect of this paper is that, we have done a significant longitudinal study comprising data for ten long years (2012–2021) across all the nine different member states to test our hypothesis. This is a significant contribution of this paper which add to the existing studies in this domain.

This paper is organized into six major sections which focus on literature review, research method, results discussion, academic and managerial implications, limitations and future research directions. Finally, the paper ends with a summary observation in the conclusion section.

2 Literature Review

Earlier studies about innovation have tried to analyze the impact of various inputs factors affecting innovation outcome. Patents have been one of the significant proxies for measuring innovation outcome globally. In many countries, various researchers have tried to measure causality relationship between innovation represented by patent filings, R&D investment, R&D funding and econometric growth factors including GDP, productivity among others. There are other studies too which try to understand the impact of institutional ecosystem specifically regulations, etc. on the innovation activities in various countries.

In the extant literature, most studies can be found trying to analyze the relationship between innovations with patent data used as a proxy for innovation to analyze economic growth where per capita GDP is used as a proxy for measuring economic growth. Crosby [2], in his paper, has used patent data as a measure of innovation to analyze its impact on economic growth represented by real GDP in Australia. In his empirical study, it can be found that in the short-term relationship between patent and GDP may be negative.

In his paper, Blind [5] tries to differentiate between economic, social and institutional regulations following the OECD taxonomy on regulations, and their impact on innovation activities. Here, the author aims to introduce a comprehensive and comparative approach to quantitatively investigate the innovation impacts in 21 OECD countries for the period of seven years (1998–2004). Among other conclusions, this empirical study confirms the hypothesis that modifications in selected regulatory and legal framework conditions have a significant influence on the dynamics of the innovative performance of OECD countries measured by the intensity of world patent applications. Here, we find that the author focuses on OECD countries only on limited institutional factors. Hence, there is a significant motivation to understand these phenomena in the context of other countries and specifically the G20-MIC member countries.

Bayarçelik and Taşel [1], in their paper, examine impact of innovation on R&D-driven growth models, where they use patent data as one of the measures of innovation. In this study, the researchers try to analyze the relationship between key markers of innovations represented by patents, R&D expenditure, and R&D employees with economic growth represented by GDP. Their study showed a significantly positive relationship between R&D investment and R&D employee with GDP. However, this empirical study also showed significantly negative relationship between number of patents and GDP.

On the contrary, results from a different study by Zachariadis [10] who tries to analyze whether growth is induced by R&D show that there exists a positive relationship between R&D and economic growth. In this empirical research, Zachariadis [10] uses patent data and R&D expenditure as proxies of innovation, gross output and productivity growth are used to measure economic outcome. This empirical study done by Zachariadis [10] provides evidence of positive relationship among R&D expenditures, patenting and productivity.

In another empirical study related to invention, innovation and economic growth for Japanese and South Korean economies, the author, Sinha [6] tries to establish causality between patents as a proxy for innovation and GDP as measure of economic growth. In case of Japan, this study is able to establish a long-term positive impact between real GPD and patents. At the same, a two-way causality between real GDP and patent growth is also established. However, in case of South Korea, it is interesting to note that it is difficult to establish significant cointegration or causality between real GDP and patent growth. Some evidence of real GDP positively impacting patent growth is seen; however, the reverse causality cannot be evidenced.

While we find many studies in extant literature related to innovation, patents and GDP are able to identify causality, there are other studies which based on multiple country data try to classify patents into resident and non-resident patents and try to analyze trends related to GDP growth by conducting various regression analysis. One such research paper by Mirzadeh and Nikzad [11] examines the trend between resident patents, non-resident patent applications (as the independent variables) and the country GDP (as the dependent variable). In this paper, the authors try to present the best trend by conducting various regression tests (exponential, linear, logarithmic, polynomial) and analyzing the regression results (R^2 value).

In a different study [4], the authors focus on various socio-economic factors which stimulate innovation in the European Union (EU). This paper examines a multitude of factors related to innovation ecosystem and not necessarily macro-economic factors. The authors use multiple regression models to investigate their hypothesis and present regression data to explain the causality among the dependent and independent variables. This study concludes that private and public R&D stimulates patent filing which is considered as a proxy for measuring innovation performance.

In his paper, Blind [5] tries to differentiate between economic, social and institutional regulations following the OECD taxonomy on regulations. Here, the author aims to introduce a comprehensive and comparative approach to quantitatively investigate the innovation impacts in 21 OECD countries for the period of seven years (1998 to 2004). Among other conclusions, this empirical study confirms the hypothesis that modifications in selected regulatory and legal framework conditions have a significant influence on the dynamics of the innovative performance of OECD countries measured by the intensity of world patent applications. Here also, we find that the author focuses on OECD countries only.

In summary, our literature review points to two critical research gaps which provide us with an opportunity to conduct an empirical study based on ten years panel data from 2012 to 2021 and quantitatively analyze the impact of institutional ecosystem factors on innovation activities within the nine geo-politically important middle-income member countries of the G20 block or the G20-MIC countries [8].

3 Research Question and Hypotheses Formulation

Based on the literature review, we understand that feel the need to understand the phenomena of innovation measured in terms of patent propensity, specifically in the G20-MIC countries and analyze the implications of various changes in the institutional ecosystem as well as GDP on the patent propensity. Hence, we ask two fundamental research questions and accordingly postulate the following hypotheses to investigate this research phenomena. Firstly, is there a causal relationship between domestic institutional ecosystem and innovation propensity of firm's measures as total patents filed in the G20-MIC countries. We also want to understand if other commonly used macro-economic factors like GDP impact innovation activities in the same landscape of all 9 G20-MIC member countries. Secondly, the next research question is: does change in GDP impact patent filing in the G20-MIC member countries. We hypothesize that the number of patents filed in the countries under study is positively impacted by the changes in the institutional ecosystem. We also hypothesize that patents filed in the countries under our study are also positively impacted by the changes to the GDP of these countries. So we formulate the following hypotheses to investigate our research questions:

Hypothesis 1: H_1

Patents filed in a G20-MIC country is positively impacted by domestic institutional ecosystem changes.

Hypothesis 2: H_2

Patents filed in a G20-MIC country is positively impacted by GDP changes.

4 Research Methodology

The main purpose of this empirical study is to examine and analyze the causal relationship between institutional ecosystem, GDP changes and patents filed in the nine different G20-MIC countries. We also try to investigate causal relationship especially reverse causal relationship between patent data and GDP of G20-MIC countries as has been done in other developed countries like Japan and South Korea [6]. In extant literature, we find various authors have used quantitative methods and specifically regression modeling technique to analyze the casual relationship. For a similar study in the European Union, authors [4] have used multi-linear regression modeling to understand the impact of various factors on innovation outcome. Hence in this study we also use this tried and tested regression modeling and multiple linear regression execution as done by Carvalho et al. [4] for our hypothesis testing. We also conduct various tests to ensure the research data used for this study is statistically appropriate. Consistent with earlier quantitative studies, in this study, we have used established statistical tests to validate data appropriateness and to remove any biases during our research. Tests for normality (using probability plot, histogram), multi-collinearity (VIF) and auto-correlation (DW) are also done to remove biases in data. Results of these tests are discussed in the analysis section of this paper. In Table 1, we provide a list of all the nine middle-income countries in G20 which are part of this research study.

Table 1 Middle-income countries in G20 group of nations (G20-MIC)

S. No.	G20 countries	Acronym	G20 status	World Bank category	Notation
1	Argentina	AG	Member	Middle-income economy	G20-MIC
2	Australia	AU	Member		
3	Brazil	BZ	Member	Middle-income economy	G20-MIC
4	Canada	CN	Member		
5	China	CH	Member	Middle-income economy	G20-MIC
6	France	FR	Member		
7	Germany	GR	Member		
8	Japan	JP	Member		
9	India	IN	Member	Middle-income economy	G20-MIC
10	Indonesia	ID	Member	Middle-income economy	G20-MIC
11	Italy	IT	Member		
12	Mexico	MX	Member	Middle-income economy	G20-MIC
13	Russia	RU	Member	Middle-income economy	G20-MIC
14	South Africa	RSA	Member	Middle-income economy	G20-MIC
15	Saudi Arabia	SA	Member		
16	South Korea	SK	Member		
17	Turkey	TK	Member	Middle-income economy	G20-MIC
18	UK	UK	Member		
19	USA	US	Member		
20	EU	EU	Member		
21	Spain	SP	Permanent guest		

Table 2 Dependent and independent variables

Data sources/variables	Acronym	Source
Change in total patents filed per million population	ΔRNRPATPMP%[a]	WIPO, World Bank
Change in GDP per million population	ΔGDPPMP%[b]	WIPO, World Bank
Total patent rights limitations percentage score	PATRIGHTSLIM%[b]	GIPC[c]
Total institutional requirements percentage score	PATREQCII%[b]	GIPC[c]

[a] Dependent variable
[b] Independent variables
[c] Global Innovation Policy Center

5 Dependent and Independent Variables

For hypothesis testing, we defined the dependent and independent variables. Change in total patents filed per million population in a year is the dependent variable used for this study. In our research, we have introduced two independent variables "Total patent rights limitations percentage score" and "Total institutional requirements percentage score" which represent the overall institutional ecosystem in the emerging economies. These two variables were calculated from annual global innovation index reports of the US Chamber International IP Index [12] from 2012 till 2021.

Extensive datasets related to patent filing by resident and non-resident citizens, national population and GDP data for 9 G20-MIC countries for ten years (2012–2021) were collected and analyzed. The dependent and independent variables and their sources are mentioned in Table 2.

6 Data Collection

For our empirical study, we collected data from various secondary sources. We focused on reliable sources like the World Intellectual Property Organization (WIPO), the World Bank and also the Global Intellectual Property Center (GIPC). We started our empirical study by collecting data for the dependent and independent variables from the secondary sources.

Initially from the World Intellectual Property Organization [13, 14], we were able to get ten years data (2012–2021) related to our dependent variable "ΔRNRPATPMP%" which represented patent filing and for the independent variable "ΔGDPPMP%" representing GDP change for all the nine G20-MIC countries under study. For the same ten years period, population data was also collected for all the 9 G20-MIC countries under study from the World Bank database. So, our initial sample dataset for this empirical study had a sample size $N = 90$. For our key independent variables "PATRIGHTSLIM%" and "PATREQCII%" which represent the institutional ecosystem changes, we collected data from annual global innovation

Table 3 Dataset and data source

Datasets	Source
Total resident patents filed per country per year	WIPO[a]
Total non-resident patents filed per country per year	WIPO
GDP per country per year	WIPO
Population per country per year	World Bank

[a] World Intellectual Property Organization

index reports of the US Chamber International IP Index [12] starting from 2012 till 2021. We initially planned to also consider data for years 2010 and 2011; however, since no data was available for key independent variables "PATRIGHTSLIM%" and "PATREQCII%" for these years, we dropped the years 2010 and 2011 in our final study. Table 3 provides the summary of the secondary datasets data sources used for our research.

7 Analysis and Discussion

The following section provides a detailed analysis of the hypothesis testing conducted across the mentioned datasets and the collected panel data. After collating the relevant data for all the secondary datasets, data screening was conducted to remove missing data and significant outliers, if any. This data screening resulted in a final dataset of sample size of $N = 59$ for our empirical study representing all the dependent and independent variables over ten years from 2012 through 2021 for all 9 G20-MIC countries under study. Institutional rights, limitations, institutional requirements were seen to vary significantly across the nine different G20-MIC countries under study.

8 Regression Modeling

The main purpose of our study is to identify the best fit regression model which explains the relationship between the dependent variable "ΔRNRPATPMP%" and the three independent variables as mentioned in Table 2. So, we use multiple linear regression approach where the independent variables are introduced step-wise. The sequence of entering the independent variable also matters. Since we are mainly interested to understand the causality relationship between institutional ecosystem changes and how it impacts the dependent variable, the independent variables are entered into the regression equation in the order sequence "PATRIGHTSLIM%" followed by "PATREQCII%" followed by "ΔGDPPMP%." So, using the independent variables, we are able to design three models, and the model summary results are presented in Table 4.

Table 4 Model summary results to identify best fit model

Model	Independent variable	R	R^2	Adj. R^2	Std. error of the estimate	Durbin–Watson
1	PATRIGHTSLIM%	0.379	0.144	0.129	9.864	1.573
2	PATRIGHTSLIM% PATREQCII%	0.392	0.154	0.123	9.894	1.564
3	PATRIGHTSLIM% PATREQCII% ΔGDPPMP%	0.569	0.323	0.286	8.927	1.938

Note. ΔRNRPATPMP% is the dependent variable

As we can see from the summary results in Table 4, R^2 as well as the adjusted R^2 value of Model 3 is the highest hence confirming that Model 3 is the best fit model for this regression analysis. It can be also seen that for Model 3, the DW statistic (Durbin–Watson) is the closest to 2 which suggest that this model does not suffer from auto-correlation. So, the linear regression model for our empirical study can be represented by Eq. 1:

$$\Delta\text{RNRPATPMP\%} = \alpha + \beta_1 \cdot (\text{PATRIGHTSLIM\%})_i + \beta_2 \cdot (\text{PATREQCII\%})_i + \beta_3 \cdot (\Delta\text{GDPPMP\%})_i + \varepsilon_i \quad (1)$$

9 Regression Execution: Tests for Normality, Multi-collinearity and Auto-correlation

As presented in an earlier research work by Carvalho et al. [4], multi-linear regression is a proven technique for such type of quantitative study. Therefore, in our study, we follow a similar research methodology [4], and multi-linear regression analysis is conducted to understand the causality relationship between the dependent variable and the independent variables. The regression equation is given in Eq. (1), and regression tests are conducted across various datasets to validate this linear relationship among the dependent and independent variables. While executing the linear regression, test for normality, multi-collinearity and auto-correlation was also considered to validate data appropriateness and to remove any biases. We used histogram to validate normality test as can be seen in Fig. 1.

During our regression analysis Durbin–Watson (DW) statistic was close to 2, and Variance Inflation Factor (VIF) was observed to be less than 3, thereby validating absence of auto-correlation and absence of multi-collinearity in the dataset, respectively. DW and VIF statistic are shown in Figs. 2 and 3. Descriptive statistics for all the variables are presented in Fig. 4.

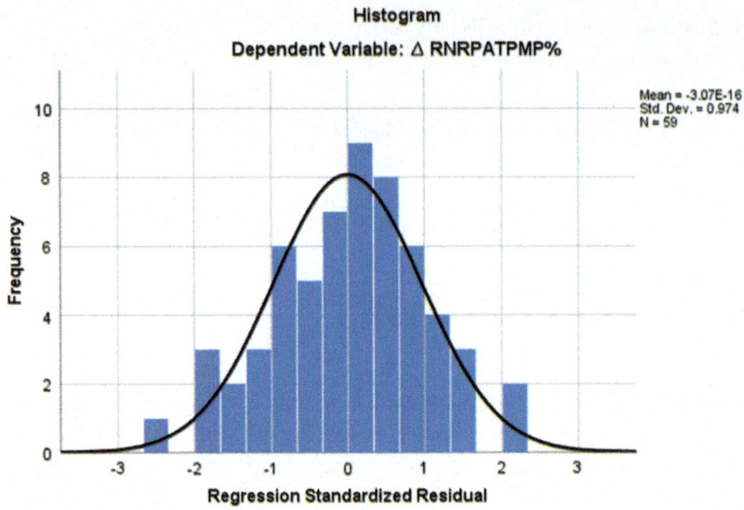

Fig. 1 Normality test

VIF Statistics for the Independent Variables[a]

Model		Unstandardized Coefficients B	Unstandardized Coefficients Std. Error	Standardized Coefficients Beta	t	Sig.	Collinearity Statistics Tolerance	Collinearity Statistics VIF
3	(Constant)	-6.006	2.722		-2.207	.032		
	PATRIGHTSLIM%	.206	.121	.320	1.703	.094	.348	2.875
	PATREQCII%	-.051	.091	-.103	-.565	.575	.372	2.690
	Δ GDPPMP%	1.267	.341	.437	3.715	.000	.890	1.123

a. Dependent Variable: Δ RNRPATPMP%

Fig. 2 Multi-collinearity test

DW Statistic for Regression Model 3[b]

Model	R	R Square	Adjusted R Square	Std. Error of the Estimate	Durbin-Watson
3	.569[a]	.323	.286	8.926759420	1.938

a. Predictors: (Constant), Δ GDPPMP%, PATREQCII%, PATRIGHTSLIM%

b. Dependent Variable: Δ RNRPATPMP%

Fig. 3 Auto-correlation test

10 Results and Discussion

After the multi-linear regression execution is completed, we analyze the test results and the estimated regression coefficients values to find significance levels of each

Summary Statistics of Variables

	N Valid	N Missing	Mean	Std. Deviation	Minimum	Maximum
Δ RNRPATPMP%	59	0	2.402607572	10.56772935	-16.3152656	40.77415250
PATRIGHTSLIM%	59	0	32.81955811	16.40324861	12.50000000	68.75000000
PATREQCII%	59	0	25.2119	21.11411	.00	75.00
Δ GDPPMP%	59	0	2.315086195	3.642487707	-4.35135224	14.61794927

Fig. 4 Summary statistics

Table 5 Estimated regression coefficient and significance levels

Model		Unstandardized coefficients		Standardized coefficients	*t*	Sig.
		B	Std. error	Beta		
3	(Constant)	−6.006	2.722	–	−2.207	0.032**
	PATRIGHTSLIM%	0.206	0.121	0.320	1.703	0.094*
	PATREQCII%	−0.051	0.091	−0.103	−0.565	0.575
	ΔGDPPMP%	1.267	0.341	0.437	3.715	0.000***

Note. Significance levels *** $p < 0.01$; ** $p < 0.05$; * $p < 0.1$ and ΔRNRPATPMP% is the dependent variable

independent variable in the regression model. In Table 5, results of the linear regression for the estimated coefficient are represented. From our tests, we conclude that among the various regression models used to test the causality, regression model 3, was found to be the best fit regression model to explain the causality relationship between the dependent and independent variables used in this empirical study. Hence, we focus our attention on regression model 3.

From Table 5, we infer that our dependent variable "ΔRNRPATPMP%" is positively impacted by changes in independent variables "PATRIGHTSLIM%" and "ΔGDPPMP%." We also infer that both the independent variables are statistically significant while they positively impact the dependent variable "ΔRNRPATPMP%." We can also infer that although both "PATRIGHTSLIM%" and "ΔGDPPMP%" positively impact the dependent variable; however, "ΔGDPPMP%" impacts the dependent variable the most. However, it can also be seen that the independent variable "PATREQCII%" is not significant and hence seems to not impact our dependent variable "ΔRNRPATPMP%."

Hence from our hypothesis testing and the results of our significance tests as summarized in Table 5, we can conclude that both of our hypotheses (H_1 & H_2) prove to be correct, and hence, they hold good. Both the independent variables (changes in institutional ecosystem and changes in GDP) are statistically significant, and results show that they positively impact the dependent variable. Equation (1) can now be modified as per the estimated regressions coefficient given in Table 5. The new regression model is given in Eq. (2):

$$
\begin{aligned}
\Delta\text{RNRPATPMP\%} = {} & -6.006 + 0.206(\text{PATRIGHTSLIM\%})_i \\
& - 0.051(\text{PATREQCII\%})_i + 1.267(\Delta\text{GDPPMP\%})_i \\
& + \epsilon_i
\end{aligned} \tag{2}
$$

11 Academic and Managerial Implications

This empirical study touches upon one of the most researched topics in the field of innovation and economic growth. It takes into consideration various researchers conducted by various researchers around the globe to investigate the interrelationship between innovation and economic growth. Traditionally patent data has been used as a proxy for innovation and hence tremendous research has been done related to patent data and various other factors that depend on patent data. However, there still exists gaps in research especially in the context of emerging economies where enough research data is not available to understand patent related econometric growth impact. Hence, our current study is a step toward bridging some of the research gaps related to patent data and understand its causality with respect to governance and policies enabling patent ecosystem especially in the emerging global economies.

From an academic standpoint, this empirical work with add to the body of knowledge related to intellectual property research and development. This research will also help other scholars to extend this work with more global datasets from various other economies.

From a managerial standpoint, results of this research can help government institutions leaders, policy makers, innovators in R&D organizations and government institutions in the G20-MIC countries to contribute to foster better institutional ecosystem through more awareness and employee engagement related to rights and limitation of the patent regulations in particular. As our research suggests, this will also enhance the innovation propensity of firms and enhance number of patent filing in these countries.

12 Limitations and Future Research Directions

Firstly, in this study, we used extensive data from secondary sources which were publicly available from reliable sources across ten years (2012–2021) involving nine emerging economies. However, in future, we feel that patent data can be fetched from other paid sources, for better correlation with public sources. Hence, reliance on only publicly available patent repository can be considered a limitation of this research work.

Secondly, while we focus on middle-income countries in the G20 block, the general applicability of our findings to other middle-income countries outside G20 may not be appropriate due to socio-economic-demographic variances. Hence, this limitation can be used to enhance future studies.

Finally, we have omitted institutional factors related to computer implemented innovations (CIIs). This is because, currently CII related data is not available in plenty in the public domain. However as digital technologies become more and more common place, innovation would center around CIIs. Hence, for future researchers, it is pertinent to focus on CIIs and try to investigate their impact on innovation activities, especially in the emerging economies.

13 Conclusion

In this paper, we have addressed two critical research gaps in extant literature in the innovation strategy domain. It was observed that most researchers focused on patent data as a proxy for innovation, and investigated a causality against various growth factors especially GDP and limited number of institutional factors. However, very little was done to understand the importance of institutional ecosystem as a whole which impacted innovation activities measured by patent filing by inventors in the middle-income countries under study. Hence, a need was felt to investigate this perspective of innovation research. Apart from the commonly used independent variable, i.e., GDP, two new independent variables were chosen to represent the institutional ecosystem. As most emerging economies lag the developed countries especially in the field innovation, it was all the more important to investigate how changes to institutional ecosystem may impact patent filing and thereby enhance the intellectual property of the emerging economies, specifically all the nine middle-income economies which have significant geo-political and economic clout as members of the G20 block. With data collected from extremely reliable secondary sources for a period of ten years for nine different emerging countries, quantitative research techniques were used to pursue this empirical study. Multiple linear regression analysis was done, which validated our hypothesis that change in institutional ecosystem and change in GDP positively impacts innovation propensity and thus increase patent filing activity by resident and non-resident citizens in these countries. As digital technologies drive innovations in the contemporary times, we suggest future researchers should extend this research to include computer implemented inventions (CIIs) while analyzing their impact on innovation activities in various countries.

References

1. Bayarçelik, E.B., Taşel, F.: Research and development: source of economic growth. Procedia Soc. Behav. Sci. **58**, 744–753 (2012). https://doi.org/10.1016/j.sbspro.2012.09.1052
2. Crosby, M.: Patents, innovation and growth. Econ. Rec. **76**(234), 255–262 (2000). https://doi.org/10.1111/j.1475-4932.2000.tb00021.x

3. Peeters, C., Pottelsberghe de la Potterie, B.V.: Innovation strategy and the patenting behavior of firms. In: Innovation, Industrial Dynamics and Structural Transformation, pp. 345–371. Springer, Berlin, Heidelberg (2007)
4. Carvalho, N., Carvalho, L., Nunes, S.: A methodology to measure innovation in European Union through the national innovation system. Int. J. Innov. Reg. Dev. **6**(2), 159 (2015). https://doi.org/10.1504/ijird.2015.069703
5. Blind, K.: The influence of regulations on innovation: a quantitative assessment for OECD countries. Res. Policy **41**(2), 391–400 (2012)
6. Sinha, D.: South Korea: evidence from individual country and panel. Appl. Econ. Int. Dev. **8**(1), 181–188 (2008)
7. Hanusch, H., Chakraborty, L., Khurana, S.: Fiscal policy, economic growth and innovation: an analysis of G20 countries. Levy Economics Institute, Working Paper No. 883 (2017)
8. World Bank: World Bank Population Data (2015). http://data.worldbank.org/indicator/SP.POP.TOTL?locations=ET
9. India, I.P.: Guidelines for examination of computer related inventions (CRIs), pp. 1–18 (2017). http://www.ipindia.nic.in/writereaddata/Portal/Images/pdf/Revised__Guidelines_for_Examination_of_Computer-related_Inventions_CRI__.pdf
10. Zachariadis, M.: R&D, innovation, and technological progress: a test of the Schumpeterian framework without scale effects. Can. J. Econ./Rev. Can. Econ. **36**(3), 566–586 (2003)
11. Mirzadeh, A., Nikzad, N.: An analysis of relation between resident and non-resident patents and gross domestic product: studying 20 countries. Int. J. Manag. Soc. Sci. **1**(2), 26–29 (2013)
12. GIPC: U.S Chamber International IP Index. February, p. 241 (2019). https://www.theglobalipcenter.com/
13. Wipo: World intellectual property indicators 2010. In: World Intellectual Property Organization, vol. 1 (2010). http://www.wipo.int/export/sites/www/freepublications/en/intproperty/941/wipo_pub_941_2013.pdf
14. World Intellectual Property Organization: WIPO IP facts and figures. In: WIPO Economics and Statistic Series (2016). https://www.wipo.int/edocs/pubdocs/en/wipo_pub_943_2020.pdf

Unique and Secure Account Management System Using CNN and Blockchain Technology

Kumar Priyanka, S. Skandan, S. Shakthi Saravanan, Ranjit Chandramohanan, M. Darshan, and S. R. Raswanth

1 Introduction

The Internet has become the main source of communication, information and entertainment these days. With the boom in the online presence due to the COVID-19 pandemic, the need for a secure and reliable account management system also grew. Video streaming service providers such as Netflix and Amazon provide users with an option to create a new account to get a free trial. Some users exploit this service by creating multiple fake accounts. In social media Web sites like Instagram and Snapchat, there is an abundance of fake user accounts where these fake user accounts masquerade and this also leads to widespread misinformation across the internet. On the other hand, service providers collect users personal information and keep it on their private database, which might be a concern to the users. The service providers might misuse or share these private details with any third-party companies

K. Priyanka (✉) · S. Skandan · S. Shakthi Saravanan · R. Chandramohanan · M. Darshan · S. R. Raswanth
Department of Computer Science and Engineering, Amrita School of Engineering, Amrita Vishwa Vidyapeetham, Coimbatore, India
e-mail: k_priyanka@cb.amrita.edu

S. Skandan
e-mail: cb.en.u4cse18374@cb.students.amrita.edu

S. Shakthi Saravanan
e-mail: cb.en.u4cse18355@cb.students.amrita.edu

R. Chandramohanan
e-mail: cb.en.u4cse18369@cb.students.amrita.edu

M. Darshan
e-mail: cb.en.u4cse19126@cb.students.amrita.edu

S. R. Raswanth
e-mail: cb.en.u4cse19648@cb.students.amrita.edu

J. Singh et al. (eds.), *Mobile Application Development: Practice and Experience*, Studies in Systems, Decision and Control 452,
https://doi.org/10.1007/978-981-19-6893-8_11

or advertisers, and this affects the users that are not comfortable with their private information being spread on the Internet. Hackers can also attack these private companies that store this information and leak it to the public, which leads to security compromise. Our proposed idea addresses these issues and aims to form a secure and unique account management system.

2 Literature Survey

Lim et al. [1]. The conventional means of logging in with an email and a password have been notoriously known as an easy way for hackers into users' accounts. Every online service provider has its variant of providing account management and authentication. A unique and secure account management system should aim to prevent unauthorized access of user credentials from third-party service providers. The paper brings to notice this idea of a decentralized network of all client credentials secured under the layer of a blockchain network. Issues of this paper discuss the integration of this idea with traditional authentication methods.

El Haddouti and El Kettani [2]. The paper discusses implementing an account and identity management system using existing blockchain technologies. It compares strategies based on technology's means of providing users with a decentralized identity on the blockchain with a secure verification of account login. The paper addresses the issue of the protection of stored user credentials and data being unclear or unexplained in some technologies, along with privacy requirements affecting existing blockchain applications.

Chen and Jenkins [3] study the performances of PCA and traditional machine learning algorithms such as LDM, SVM and KNN. PCA is used to reduce the size of each image by extracting the linear combinations of the eigenvalues of the original image to form a dataset of eigenfaces. The authors in the paper claim that this improves the speed and accuracy of the existing methods to do such tasks. In the paper, they use the ORL dataset which contains 10 images of 40 individuals in different settings. The stages which they flow to arrive at their result are PCA processing, split dataset into test and training sets and then applying the machine learning models. From the experiments they conduct, they conclude that SVM looks to be the best choice if there is little sensitivity to running speed and a significant need to improve identification accuracy. KNN, on the other hand, could be able to strike a better compromise between running speed and identification accuracy.

Finizola et al. [4] do a comparative study between traditional machine learning face recognition algorithms and deep learning-based face recognition algorithms. They initially stated that the main reason for their study is to prove that deep learning models are better than traditional machine learning models. In the paper, they use KNN, OPF, SVM, extreme learning machine, ANN, as representatives of traditional machine learning models and CNN model, autoencoder model with ELM classifier, as representatives of deep learning models. The datasets used are JAFFE, YALE, AR and SDUMLA. Jonnathann and his team use K-fold cross-validation methods (tenfolds) for training both the traditional ML models and the deep learning models. In the paper, they use performance measurements such as Euclidean distance, p-value, null hypothesis, to arrive at the following conclusion. The traditional models performed better on the YALE dataset and AR dataset, and the deep learning models performed well in the SDUMLA dataset. With the use of feature extractors, the traditional models performed better on datasets with fewer elements. The paper mentions that the understanding of deep learning models is still in infancy, and future work on them would improve the performance of deep learning models.

Bakre et al. [5] explain the four phases of identity: centralized identity, federated identity, user-centric identity and self-sovereign identity. Self-sovereign identity is the latest and most secure identity management technique. In it, the end-users have total control over how to share their identity and to whom to share it. According to this paper, the concept of self-sovereign identity involves the usage of blockchain, secure sockets layer (SSL), single sign-on (SSO) and Kerberos. Blockchain technology is used as it makes the database decentralized, transparent and fully secure. SSL is to be used for creating a secure channel to the internet and Web site. SSO is used to make the user log in to multiple applications with a single successful login attempt. So with all these working together, the goal of the paper was to propose a secure login system that uses blockchain as the database. The outcome of our survey on this paper was that it was only a research paper and the idea was not implemented. It also required the service vendors to implement it to their Web site and was not relying on a universal API system.

Cao et al. [6] discussed ways to overcome common obstacles with any deep learning task. The gathering of datasets is a must for any convolutional neural network to operate, and this study fulfills that need. The collection includes a wide range of pictures with varying poses, ages, lighting, ethnicity and occupations. Various models were also utilized to evaluate the increased performance. Finally, their study demonstrated state-of-the-art performance on the IJB-A and IJB-B face recognition benchmarks, significantly outperforming the prior benchmark. The findings of this study established the groundwork for the dataset that would be used in the suggested solution.

Coskun et al. [7] concluded that two normalization procedures were added to two of the layers in their study effort, resulting in a modified convolutional neural network design. Batch normalization resulted in faster network findings. In the fully connected layer of CNN; distinct facial characteristics were discovered, and SoftMax classifiers were utilized to categorize faces. When the model was put to the test, it produced better results and performance.

3 Proposed System

See Fig. 1.

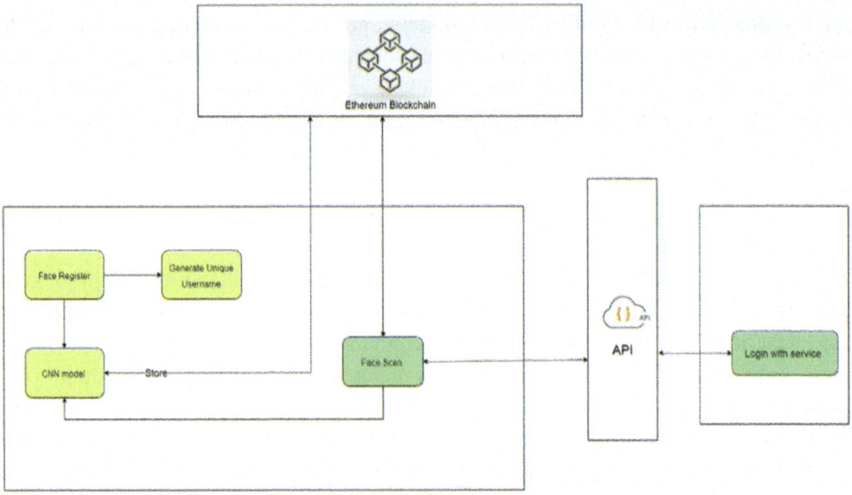

Fig. 1 Architecture diagram

3.1 *Front-End Account Creation*

The front end of the account creation service is created using ReactJS and web3.js (for blockchain connectivity). The user first clicks to create an account option on the Web site. The user is then prompted to enter his unique ID, and then, the user is also prompted to store his facial data, so that they can log in to their linked Web sites using their face. During the login process (redirected from other Web sites), the user can log in with his username and facial data.

3.2 *Face Recognition Model*

Here, we make use of the VGG16 model, which is a convolutional neural network model used for the classification of faces accurately. For adding face data, we take a different set of images of the user for training. This data is stored in a centralized database. For training, we separate new faces into different classes (folders) with their names. We modify the final layer of the VGG16 model to n-different classes available and create the model with the weight of ImageNet. We then fit the model with the train data of the different classes (faces). For validation of if the faces were tested also with a new set of face images. Once the validation is done, we send back an acknowledgement to the account login in the blockchain.

3.3 Blockchain Storage

The blockchain is used as a storage system. The blockchain is implemented on the Ethereum test net and a local blockchain is created using Ganache. The smart contracts will be written using the programming language Solidity and tested using Truffle Suite. After the user creates an account, his account credentials are stored on the blockchain in a map structure. When a user is redirected to the API service during the login process, the data is checked on the map, and if it matches, the login is successful.

4 Implementation and Validation

4.1 Blockchain Module

- **Implementation**: Ethereum's blockchain supports smart contracts, which allow developers to write code that exists on the blockchain and run it when certain conditions are met. Solidity is the programming language used to write smart contracts on the Ethereum network. The smart contract that is written currently allows the users to create an account using a name and the output of the facial recognition model, face descriptor values. After that registration step, the users can enter the same details and perform the login authentication process. If the details match with the registered details, a popup is displayed which says "logging in." If the details do not match, an error popup is displayed (Fig. 2a). The contract *AccountManagement* consists of two functions: *createAccount* and *checkLogin*. The account creation function takes in a name and an ID as arguments and stores

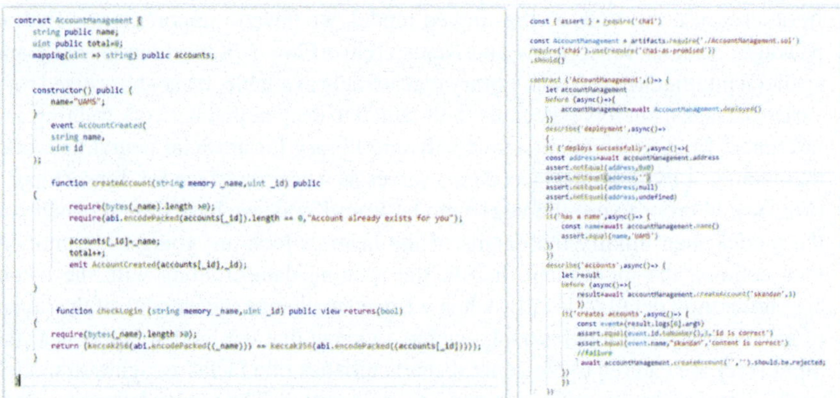

Fig. 2 **a** Ethereum smart contract, **b** smart contract validation testing code

them in the hashmap "accounts." The map maps an ID to a name. There also exists a required condition, which makes it impossible for the same ID to exist again on the map. This blocks a person from creating multiple new accounts. The registration function has gas fees as it has to modify the blockchain. The function *checkLogin* also takes the same arguments as the *createAccount* function. It is a view-only function that does not modify the blockchain, and so there is not any gas fee involved when it is called. The function just returns the Boolean value true if the ID maps to the same name in the "accounts" map; otherwise, it returns the Boolean value false.

- **Validation**: (Fig. 2b). For the validation of the smart contract, Truffle Suite is being used. chai is a JavaScript library that allows for assertions to be made. Using asserts and predefined inputs, we describe two test conditions. The first test "deployment" checks if the smart contract properly exists on the blockchain. The second test "accounts" tests if the smart contract function *createAccount* works as expected, both in the case of correct and incorrect inputs.

4.2 Face Recognition Module

- **Implementation**: The VGG16 model is a VGG convolutional neural network model used to classify faces correctly. VGG accepts a 224 × 224 pixel RGB picture as input. To keep the input picture size consistent for the ImageNet competition, the authors clipped out the central 224 × 224 patch in each image. We utilize a distinct collection of user photographs for training when adding face data. This information is stored in a single database. We divide fresh faces into different classes based on the user's name or ID for training purposes. We then alter the weights of the VGG16 pre-trained model's layers which lie after the bottleneck layer. The model is fitted using train data from various classes. A new set of face images is used to verify if the faces have been evaluated or not.
  In order to use the VGG16 pre-trained model, we have to import certain Python packages such as TensorFlow and Keras. TensorFlow is a machine learning and artificial intelligence software library that is freely available. It may be utilized for a variety of tasks, but it is especially well-suited to deep neural network training and inference. Keras is an open-source software library for artificial neural networks that include a Python interface. Keras serves as a user interface for TensorFlow.
  (Fig. 3a). In order to use the vgg16 model from Keras, we have to first instantiate the model, then modify the details of the layers which are going to be trained, then compile the model and then fit the training dataset along with the batch, epoch/iteration details. Usually, when we need to change the weights of the layers of an existing pre-trained model, one has to know that the hidden layers near the input layer are catered to the general characteristics of a facial recognition model, such as dimensionality reduction or feature extraction. The hidden layers nearer to the output layer are more specific to the given dataset. We need to train the layers

```
model_final.compile(loss = "categorical_crossentropy",
    optimizer = tf.keras.optimizers.SGD(lr=0.0001,
    momentum=0.9), metrics=["accuracy"])

Model: "sequential"
_________________________________________________________________
Layer (type)                 Output Shape              Param #
=================================================================
mobilenetv2_1.00_224 (Functi (None, None, None, 1280)  2257984
_________________________________________________________________
conv2d (Conv2D)              (None, None, None, 32)    368672
_________________________________________________________________
dropout (Dropout)            (None, None, None, 32)    0
_________________________________________________________________
global_average_pooling2d (Gl (None, 32)                0
_________________________________________________________________
dense (Dense)                (None, 10)                330
=================================================================
Total params: 2,626,986
Trainable params: 369,002
Non-trainable params: 2,257,984
_________________________________________________________________
```

```
checkpoint = ModelCheckpoint("vgg16_1.h5", monitor='accuracy',
    verbose=1, save_best_only=True, save_weights_only=False,
    mode='auto', period=1)
early = EarlyStopping(monitor='val_acc', min_delta=0,
    patience=40, verbose=1, mode='auto')
hist = model_final.fit_generator(generator= traindata,
    steps_per_epoch= 2, epochs= 5, validation_data= testdata,
    validation_steps=1, callbacks=[checkpoint,early])
model_final.save_weights("vgg16_1.h5")
```

Fig. 3 **a** Compiling model, **b** fitting into the model

which lie below the bottleneck layer in order to make it specific for our dataset. We then make sure that all the perfectly trained layers that are responsible for the general filtering are not trained while the output layers give feedback.
Later, we define the attributes of the layers for the model with appropriate activation functions such as ReLu or SoftMax. The layers include a 2D convolution network for feature enhancement, Dropout for removing nodes that are not contributing much value to the final output, GlobalAveragePooling2D for data size reduction by taking the average and setting the dense layer. All above are transformation layers; this is the main dense layer. The dense layer takes input from all previous nodes and gives input to all the next nodes. It is very densely connected and hence called the dense layer.
(Fig. 3b). Then, we move on to the compiling of the model, where we set the optimization algorithm, the loss function and the metrics which will be used for early stop. We will be using the categorical cross-entropy loss function which is the most common loss function for classification neural network models.

- **Validation**: In order to validate the choice of vgg16, we had initially referred to a research paper that did an in-depth analysis of the available VGG models. To bolster our choice, we implemented the MobileNetV2 model which was a model developed by Google which is a VGG variant. We did a comparison of the accuracy and the loss function of the test and train set for both the models, and the results were as follows (Fig. 4):

5 Result

5.1 *Blockchain Module*

For the current smart contract, the testing was successfully validated. The front is also linked with the blockchain, and on successful login attempts, we are given a popup with a successful message. After setting up the configuration of the truffle,

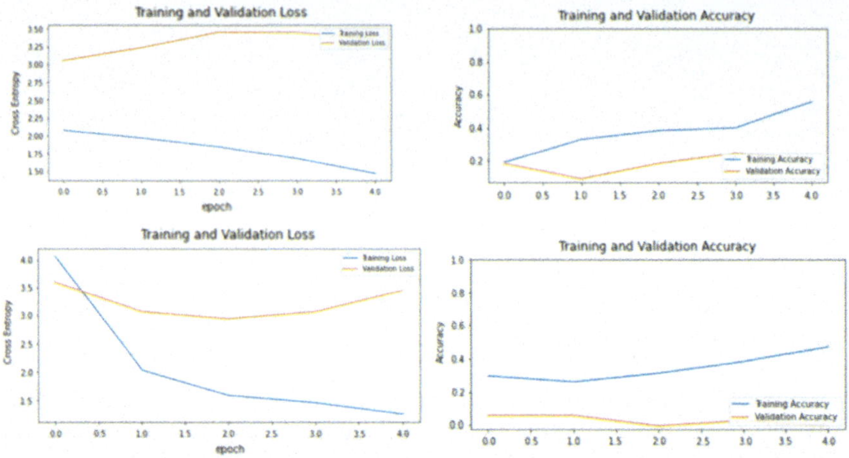

Fig. 4 **a** MobileNetV2 loss function, **b** MobileNetV2 accuracy, **c** VGG16 accuracy, **d** VGG16 loss function [Clockwise]

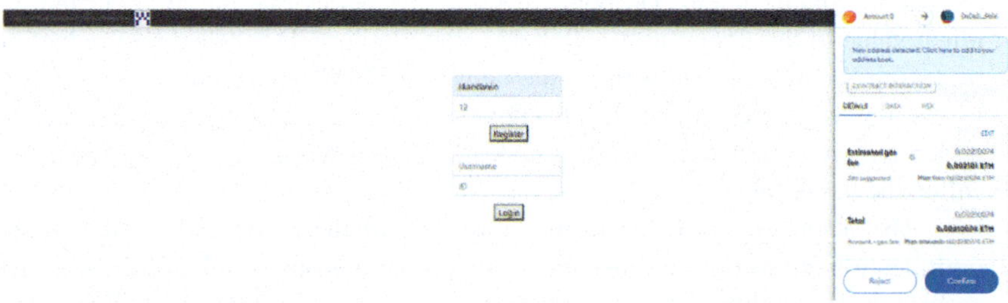

Fig. 5 Smart contract execution with interface

contracts have been deployed using the command truffle migrate. The truffle test command is used for testing the instance, which displays the passing of events and functions of the smart contract (Fig. 5).

5.2 *Face Recognition Module*

In order to test the model which, we have trained, we give to fit a set of test images to the model and get the confidence probability score for that image for each class. The trained VGG16 model has been tested with different sets of images and weights, and performance data of the model has been plotted for better understanding of the inference (Fig. 6).

```
for e,i in enumerate(os.listdir("../ORLinput/Test")):
    print(i)
    output=[]
    img = image.load_img(os.path.join("../ORLinput/Test",i),
        target_size=(224,224))
    img = np.asarray(img)
    img = np.expand_dims(img, axis=0)
    output = model_final.predict(img)
    print(output)
```

```
10_1.jpg
[[0.03050304 0.06084248 0.02795632 0.08306099 0.02367211 0.00882624
  0.7284576  0.02476986 0.00130187 0.01060944]]
14_2.jpg
[[7.5701177e-02 2.8445616e-01 5.7817411e-02 1.2315610e-01 1.9669144e-01
  6.8530053e-02 1.4903396e-01 3.3549346e-02 2.5607695e-04 1.0808286e-02]]
18_2.jpg
[[3.2148533e-02 7.0103610e-01 2.6974952e-02 6.4097732e-02 3.2956541e-02
  3.2684974e-02 6.8891600e-02 9.0948939e-03 2.1370292e-04 3.1900913e-02]]
19_2.jpg
[[1.3342737e-02 8.6660427e-01 7.7693346e-03 2.5304303e-02 3.0461648e-02
  2.4429997e-02 2.0611200e-02 6.8974257e-03 2.6669115e-04 4.3124193e-03]]
20_2.jpg
[[2.6117394e-02 6.1405796e-01 1.0857296e-02 5.7728402e-02 3.9161384e-02
  3.6433507e-02 1.3504651e-01 3.2846227e-02 9.8670826e-05 4.7652699e-02]]
```

Fig. 6 Face recognition model execution

6 Conclusion

The goal of the idea is to integrate a facial recognition module to log in into a blockchain network. The proposed approach has reinforced the facial recognition module using convolutional neural networks, which provides a secure solution to existing less efficient conventional systems. Unique and secure account management for users has been achieved using smart contracts through the Ethereum blockchain network, which serves the purpose of unauthorized access of the users' information by third-party companies and benefits the service providers using this API to solve the use of large amounts of fake accounts. As for future work, with the deep learning field ever-expanding, the implemented facial recognition module could be tuned to improve the accuracy of recognizing faces. Implementing this system on a storage-based blockchain is also another future endeavor to store other user data such as credit card information and national identity.

References

1. Lim, S.Y., Fotsing, P.T., Almasri, A., Musa, O., Kiah, M.L.M., Ang, T.F., Ismail, R.: Blockchain technology the identity management and authentication service disruptor: a survey. Int. J. Adv. Sci. Eng. Inf. Technol. **8**, 1735–1745 (2018)
2. El Haddouti, S., El Kettani, M.D.E.-C.: Analysis of identity management systems using blockchain technology. In: 2019 International Conference on Advanced Communication Technologies and Networking (CommNet). IEEE, pp. 1–7 (2019)
3. Chen, J., Jenkins, W.K.: Facial recognition with PCA and machine learning methods. In: 2017 IEEE 60th International Midwest Symposium on Circuits and Systems (MWSCAS). IEEE, pp. 973–976 (2017)
4. Finizola, J.S., Targino, J.M., Teodoro, F.G., Lima, C.A.: Comparative study between deep face, autoencoder and traditional machine learning techniques aiming at biometric facial recognition. In: 2019 International Joint Conference on Neural Networks (IJCNN). IEEE, pp. 1–8 (2019)
5. Bakre, A., Patil, N., Gupta, S.: Implementing decentralized digital identity using blockchain. Int. J. Eng. Technol. Sci. Res. **4**, 379–385 (2017)
6. Cao, Q., Shen, L., Xie, W., Parkhi, O.M., Zisserman, A.: Vggface2: a dataset for recognising faces across pose and age. In: 2018 13th IEEE International Conference on Automatic Face and Gesture Recognition (FG 2018). IEEE, pp. 67–74 (2018)
7. Cokun, M., Uçar, A., Yildirim, Ö., Demir, Y.: Face recognition based on convolutional neural network. In: 2017 International Conference on Modern Electrical and Energy Systems (MEES). IEEE, pp. 376–379 (2017)

Model for Mobile App-Based Premium Calculation for Usage-Based Insurance (UBI) of Vehicles

Sohil Pandya, Nilay M. Vaidya, Jaimin N. Undavia, Atul M. Patel, Krishna Kant, and Abhilash Shukla

1 Introduction

1.1 Artificial Intelligence

In the history of artificial intelligence, various authors have approached to define artificial intelligence with their own way, and one of them was Nilson (Ref of 1) who has define the artificial intelligence as 'Artificial Intelligence is the activity devoted to making machines intelligent and intelligence is that quality that enables and entity to function appropriately and with foresight in its environment.' It was clearly denoted by the early researchers of artificial intelligence that regardless of the particular approach artificial intelligence research has been united by from the beginning by its engagement with and mechanizing intelligence (Ref of 1 Turing). In recent times, we have witnessed rapid advancements in the field of artificial intelligence results in

S. Pandya (✉) · N. M. Vaidya · J. N. Undavia · A. M. Patel · K. Kant · A. Shukla
Smt. Chandaben Mohanbhai Patel Institute of Computer Applications (CMPICA), Charusat University, Changa, Anand, Gujarat, India
e-mail: sohilpandya.mca@charusat.ac.in

N. M. Vaidya
e-mail: nilayvaidya.mca@charusat.ac.in

J. N. Undavia
e-mail: jaiminundavia.mca@charusat.ac.in

A. M. Patel
e-mail: atulpatel.mca@charusat.ac.in

K. Kant
e-mail: krishnakant.mca@charusat.ac.in

A. Shukla
e-mail: abhilashshukla.mca@charusat.ac.in

J. Singh et al. (eds.), *Mobile Application Development: Practice and Experience*, Studies in Systems, Decision and Control 452,
https://doi.org/10.1007/978-981-19-6893-8_12

profound implications in every aspect of society and business. Any types of business weather product based or service based have enough opportunities to innovate the routine process through the advent uses of artificial intelligence. All the business owners can improve their process by incorporating the artificial intelligence in their various process like production, recruitment, competition, etc. Apart from the innovation, the artificial intelligence can be used to improve the innovative process itself which may lead to higher level of efficiency over a period of time. This may lead to dominate the direct effect in the process efficiency in business! [1]. Artificial intelligence is not only the way to improve the routine process by building the new solution from the scratch. Some applications with aids of artificial intelligence surely constitute solution with lower cost or higher quality into existing production process too. The recent stand points of artificial intelligence like machine learning, deep learning, CNN, etc., are the artifacts of such improvements in the existing systems. Such innovations have impact on large variety of business and these businesses can lead to new level of efficiency. It is also proved that the recent advances in machine learning, deep learning and neural networks through their nature to improve performance of technologies at end user point and the nature of the innovation processes are like to have a particularly large impact on innovation and growth. The incentives and economic growth of the business through the usage of such advanced technologies have motivated many researchers to shape up their solutions through development and diffusion of these technologies. Economic growth of the business may lead to use these recent tools and technologies which ultimately improve the efficiency of entire market by exponential usage of such innovative artifacts.

1.2 IOT and Robotics

Basically IoT—Internet of Things—is the network of objects (things), and the objects in the network are equipped with various sensors, software and other technologies. These equipment are used to connect the things and exchange data with other devices of systems over the Internet. The devices involved in such network ranges from ordinary household object to the most advanced industrial tools. With more than 7 billion connected IoT devices today, experts are expecting this number to grow to 10 billion by 2020 and 22 billion by 2025 [https://www.oracle.com/in/internet-of-things/what-is-iot/]. In recent times, IoT has become the most important technologies of the twenty-first century, and now, innovators are focusing on connecting every objects with each other to facilitate the routine processes. Various objects like kitchen appliances, cars, baby monitors, and thermostats are getting connected to the internet via embedded devices and which enables seamless communication made possible between people, processes, and things [2].

The technologies involved in IoT made it practical, and these technologies are listed below:

- Access to low-cost, low-power sensor technology
- Connectivity
- Cloud computing platforms
- Machine learning and analytics
- Conversational artificial intelligence.

Such recent innovations in the field of IoT encouraged all the types of industries to accommodate it at some extent. Here is the list of the some of the industries which are benefited through the advent of IoT.

- Manufacturing
- Automotive
- Transportation and logistics
- Retail
- Health care
- General safety across all the industries.

Robots, on the other hand, will play a major role in tomorrow's society, continuing to help humans accomplish many tasks, spanning assistive operations, industrial assembly, rescue management systems, military support, healthcare, and automation systems [3]. Robotics is considered as the most intelligent implementation of IoT which explores the more advanced and transformational aspects of ubiquitous connectivity among smart devices. Unlike, traditional IoT, it is not just onboard connection of different devices, but it is smarter and advanced integration of various devices at high degree of communication among them [4]. Such transformation can be achieved through robotics systems as they have in-built ability to sense, compute, take decision, and move accordingly. The synergy between IoT and robotics is depicted where intelligent devices can supervise or monitor events or processes, turn on and off the actuators, send and receive signals through various sensors and can determine a best course of action. The system can control or monitor such actions within physical world [3].

2 Literature Survey

Author name	Title	Source	Important findings
da Silveria Barreto et al. (2018)	A machine learning approach based on automotive engine data clustering for driver usage profiling classification (http://doi.org/10.5753/eniac.2018.4414)	Proceedings of XV Encontro Nacional de Inteligência Artificial e Computacional (ENIAC), Brazil (2018)	The authors proposed a model and experimented it, for driver usage profiling using machine learning approach, first by identifying labels using clustering techniques and later used classification techniques by using these labels for choosing most appropriate model. Based on the data available with authors, they found MLP is best suitable to train and test over the best partition found using KM
Soleymanian et al. (2019)	Sensor data and behavioral tracking: Does usage-based auto insurance benefit drivers? (http://doi.org/10.1287/mksc.2018.1126)	INFORMS Journal of Marketing Science (2019)	The authors studied various economical, psychological and behavioral impacts of UBI Score calculations using sensor and other data (like age, marital status, etc.), which not only helps owner/driver to have reduced insurance premiums but also helps insurance companies to predict risks, pricing strategy, and provide better value to policyholders
Soleymanian (2019)	Monitoring, sensor data, privacy and consumer behavior: the case of usage-based automobile service (Open Collection, Library, University of British Columbia)	Ph.D. Thesis, The University of British Columbia, Vancouver	In addition to the findings Soleymanian et al. (2019), other findings are: • Author have studied adoption level of UBI among different age-group in the context of privacy • Information and feedback help users to improve • Insurer's profile helps him/her to have lower insurance premiums which leads to have better/safer driving patterns, which at large reduces risk of accidents and helps society
He et al. (2018)	Profiling driver behavior for personalized insurance pricing and maximal profit (http://doi.org/10.1109/BigData.2018.8622491)	Proceedings of 2018 IEEE conference on Big Data	Authors have proposed Profile–Price–Profit (PPP) model for identifying driver behavior profiling (by proposing ensemble learning algorithm) and based on profile, a model for personalized insurance pricing (by model incorporating demographic and telematics data) which ultimately helps to maximize insurance company's profit (by proposed dynamic programming solution). Authors have considered following data of telematics: vehicular angular velocity, cool liquid temperature, acceleration, engine speed, vehicle speed
Yan et al. (2020)	Research on UBI Auto Insurance Pricing Model Based on Adaptive SAPSO to Optimize the Fuzzy Controller (http://doi.org/10.1007/s40815-019-00789-6)	International Journal on Fuzzy Systems (2020)	Authors have proposed to apply fuzzy controller and optimizing of it by adaptive SAPSO algorithm (Metropolis criterion) to find optimal fuzzy rule to ultimately identify that how this implementation can effectively and accurately determine UBI premium
Qi et al. (2020)	Scalable Decentralized Privacy-Preserving Usage-based Insurance for Vehicles (http://doi.org/10.1109/JIOT.2020.3028014)	IEEE Internet of Things Journal (2020)	Authors have implemented a unique of kind decentralized privacy preserving in UBI by proposing DUBI as a consortium blockchain system which is found more speedier and secure (by hiding identity)

(continued)

Author name	Title	Source	Important findings
Pettersson et al. (2019)	Usage-Based Auto Insurance on the Swedish Market:A Case Study (http://doi.org/10.23919/PICMET.2019.8893870)	Proceedings of 2019 Portland International Conference on Management of Engineering and Technology (PICMET)	In this paper, authors provided a case study on how UBI in Sweden was implemented as a pilot project and how it was responded among customers. It is shown that UBI has a great potential in upcoming years
Magri et al. (2019)	Determining Motor Insurance Premium in a Small Island State: The Case of Malta	International Journal of Finance, Insurance and Risk Management (2019) (http://doi.org/10.35808/ijfirm/191)	Authors have provided an insight on risk factors in determining UBI in Malta by which is also supported by survey of users and insurance companies
Narwani et al. (2020)	Categorizing Driving Patterns based on Telematics Data Using Supervised and Unsupervised Learning. Categorizing Driving Patterns based on Telematics Data Using Supervised and Unsupervised Learning (http://doi.org/10.1109/ICICCS48265.2020.9120976)	Proceedings of the IEEE International Conference on Intelligent Computing and Control Systems (ICICCS 2020)	Authors have applied Machine Learning algorithms to identify driving patterns based on acceleration and jerk (sudden break). Using unsupervised learning, they identified clusters in the context of risk
Huang et al. (2019)	Automobile insurance classification ratemaking based on telematics driving data (http://doi.org/10.1016/j.dss.2019.113156)	Decision Support Systems (2019)	Here, authors have investigated driving behavior from telematics data and improved pricing strategy for UBI by applying logistic regression and four ML algorithms (SVM, RF, XGBoost, and NN) to predict risk probability and claim frequency of insured vehicles
Cevolini et al. (2020)	From pool to profile: Social consequences of algorithmic prediction in insurance (http://doi.org/10.1177/2053951720939228)	Big data society (2020)	Authors have carefully studied social consequences in UBI and raised them by exemplifying from both insurers and insurance companies' point of view
Weidner et al. (2016)	Telematic driving profile classification in car insurance pricing (http://doi.org/10.1017/s1748499516000130)	Annals of Actuarial Science (2016)	Based on the driving profiles generated using telematics data (especially acceleration and deceleration) authors have studied risk assessment approach for UBI

3 Research Gap

1. There is lack of system which will automatically take all records.
2. There is lack of system which will continuously monitor driver regarding the safety precaution, whether they are followed by him or not.
3. There is lack of system which will take full proof action against drivers. After final profile monitoring it will make a great help to the insurance company.
4. There is no any better system available which will give the proper information so it will make a great help to the insurance company to differentiate between safe driver and unsafe driver.

5. As per the safe driving there is no any alert system which will force drivers to drive safely.
6. There is lack of observing whether the driver has drunk alcohol while driving or not.
7. The most critical part for insurance companies is to recognize the importance of environmental factor in aligning the individual risk and price.

4 Objective

1. Implementation of system which is able to track all records automatically.
2. Implementation of the system which will continuously monitor whether the safety precaution is followed by the driver or not.
3. Implementation of system which makes the differentiation between safe driver and unsafe driver with the help of vehicle telematics data.
4. Implement a system with continuous alcohol intake testing while driving the vehicle.
5. Implementation of system which will give proper information of car and driver.
6. Implementation of system which will motivate the driver to drive safely on the account of insurance.

5 Conceptual Diagram

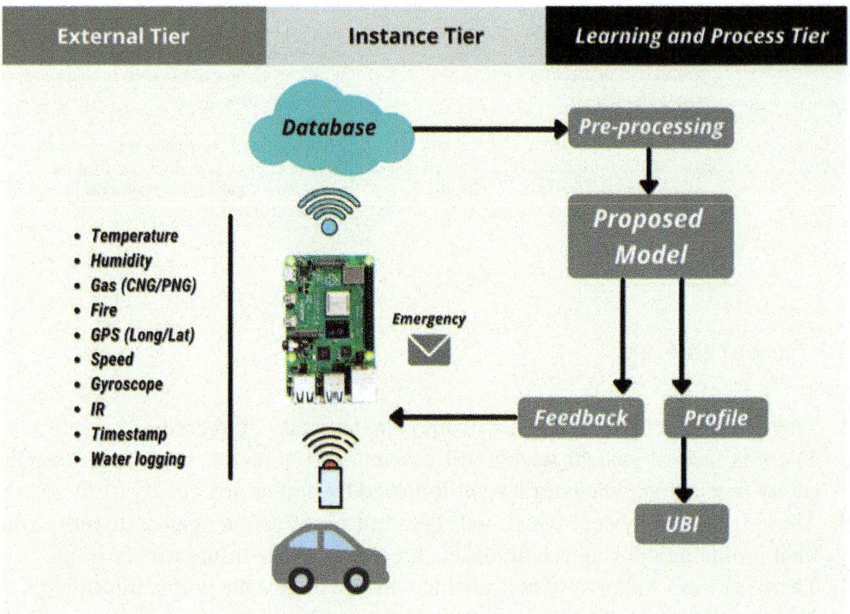

The proposed model is conceptually divided into three logical tiers:

(a) External tier
(b) Instance tier
(c) Learning and Process tier.

Each tier proposed have their distinct feature and functionalities that continuously remains active and transmits/receives the value(s) to the concerned tier. The general parameters taken into consideration so far are

1. Type of vehicle (fixed)
2. Manufacturing month and year
3. Vehicle travel reading (average travel per fixed interval)
4. Type of road pattern in which the vehicle travels
5. Speed at which the vehicle travels
    (a) Speed against the road pattern
    (b) Speed against the weather–environment
6. Service pattern
    (a) Air pressure maintained
    (b) Oil level maintained
    (c) Coolant level, and many more
7. Average weight load while traveling
8. Timestamp.

Moving vehicle, continuously captures the data about the above-mentioned criteria. On the pre-defined interval of time, instance tier transmits the data to the learning and process tier for the further processing. Learning and process tier collects, filters and further applies rules onto it to generate the informed reports which then can be communicated to the concerned tiers. The proposed model, finds the weights associated with each defined criteria. Here weights are calculated and generated from the pre-defined rules set by the appropriate domain experts. This weights are then being passed to the model for further processing. For the processing the data, we are applying multi-objective gray situation decision-making theory (MGSD) to find the current condition of a vehicle on the given criteria [5–7]. In light of these developments, this study develops a series of models capable of forecasting the understanding usage level of the vehicle by applying the gray theory and multi objective programming. Decision-making theory includes the elements like: driver/owner, event/trip/travel, parameters/criteria, vehicle condition. Here, while calculating the effect measure based on each criterion, we used weights attached. Decision-making algorithm finds the effect measures for each criteria based in their deviation of the data. With highest deviation, it uses upper limit measure which considers higher the weight values better the results. To calculate the effect measure

$$r_{ij} = u_{ij}/u_{\max},$$

where

$$u_{ij}$$

is the actual effect measure (weight) and

$$u_{\max}$$

is the highest effect measure defined by the domain expert for the criteria

$$u_{ij} \leq u_{\max} : r_{ij} \leq 1$$

With lower deviation, it uses lower limit measure which considers lower the weight values better the results. To calculate the effect measure

$$r_{ij} = u_{\min}/u_{ij},$$

where

$$u_{ij}$$

is the actual effect measure (weight) and

$$U_{\min}$$

is the lower effect measure defined by the domain expert for the criteria

$$u_{ij} \geq u_{\min}$$

$$u_{\min} : r_{ij} \geq 1$$

Decision-making theory helps in finding the effect measure based in the upper or lower criteria based in the weights and the actual values given into the model for each identified criteria. Here, there are several objectives for a given situation, so we need to find the situation decision. Situation decision based on multiple objective is referred as multiobjective situation decision-making. The effect measure and corresponding decision unit needs to be calculated by just deriving a decision matrix

$$\begin{bmatrix} r_{i1}(k) & r_{j1}(k) & r_{im}(k) \\ - & - & - \\ s_{i1} & s_{j1} & s_{im} \end{bmatrix} \quad \text{where}$$

$$r_i$$

are the effect measures for the objective, and

$$s_i$$

are the situational strategies Similarly, the multiobjective situation decision comprehensive matrix can be derived and calculated as

$$[r_{ij}{}^{\Sigma}] = 1/N \sum_{k=1}^{n} r_{ij}(k)$$

This comprehensive effect measure gives a value between 0 and 1. As it is nearer to 0, for our domain the usage is more and tentatively the premium to be calculated above the threshold value. And as the value is nearer to 1, according to the theory and algorithm, the premium to be nearer to the threshold.

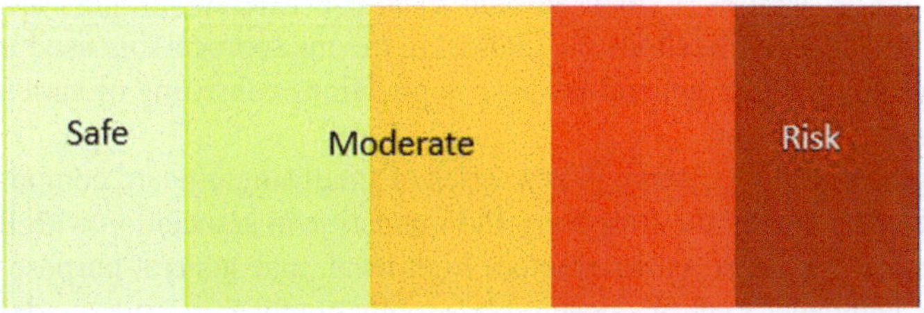

6 Conclusion

The work proposed in the chapter targets to offer usage-based premium for car insurance providers. As the usage pattern and maintenance of the car has good impact over the overall life span of the vehicle, it becomes essential to offer usage-based premium to leverage and improve the premium calculation system. To calculate this premium, eight major categories of parameters are considered with their associated weight. This weight is determined by the domain experts, and they have considered as the most affecting parameters in the given scenarios. These eight criteria are further transformed into a mathematical notation which further classifies the usage patter into three classes, which is Safe, Moderate and Risk. Based on the class label associated with each car, the insurance premium calculation is proposed. A model is proposed to collect real-time data to get rid of actual usage patter of a particular car. The proposed work is going to revolutionize the insurance industry, and also, at the same time, it will encourage the car users to drive safely with well-maintained car.

7 Important Points

1. **Usage-Based Insurance (UBI)**: Usage-Based Insurance (UBI) is a recent autoinsurance innovation that enables insurers to collect individual-level driving data, provides feedback on driving performance, and offer individually targeted price discounts based on each consumer's driving behavior.
2. **Driver's Profile**: With respect to the insurance companies.
3. **Internet of Things (IoT)**: The Internet of Things (IoT) describes the network of physical objects—'things'—that are embedded with sensors, software, and other technologies for the purpose of connecting and exchanging data with other devices and systems over the Internet.
4. **Microcontroller**: A microcontroller (MCU for microcontroller unit) is a small computer on a single metal-oxide-semiconductor (MOS) integrated circuit (IC) chip. A microcontroller contains one or more CPUs (processor cores) along with memory and programmable input/output peripherals. Program memory in the form of ferroelectric RAM, NOR flash, or OTP ROM is also often included on chip, as well as a small amount of RAM. Microcontrollers are designed for embedded applications, in contrast to the microprocessors used in personal computers or other general-purpose applications consisting of various discrete chips.
5. **Raspberry Pi**: Raspberry Pi is a series of small single-board computers developed in the UK by the Raspberry Pi Foundation in association with Broadcom.
6. **Python**: Python is an interpreted, high-level, and general-purpose programming language. Python's design philosophy emphasizes code readability with its notable use of significant whitespace. Its language constructs and object-oriented approach aim to help programmers write clear, logical code for small and large-scale projects.
7. **Cloud Storage**: Cloud storage is a model of computer data storage in which the digital data is stored in logical pools, said to be on "the cloud". The physical storage spans multiple servers (sometimes in multiple locations), and the physical environment is typically owned and managed by a hosting company.
8. **Sensors**: In the broadest definition, a sensor is a device, module, machine, or subsystem whose purpose is to detect events or changes in its environment and send the information to other electronics, frequently a computer processor.
9. **Machine Learning**: Machine learning (ML) is the study of computer algorithms that improve automatically through experience. It is seen as a subset of artificial intelligence. Machine learning algorithms build a model based on sample data, known as "training data", in order to make predictions or decisions without being explicitly programmed to do so.
10. **Cluster Analysis**: Cluster analysis or clustering is the task of grouping a set of objects in such a way that objects in the same group (called a cluster) are more similar (in some sense) to each other than to those in other groups (clusters).

11. **Classification**: Classification is the problem of identifying to which of a set of categories (sub-populations) a new observation belongs, on the basis of a training set of data containing observations (or instances) whose category membership is known
12. **Regression Analysis**: Regression analysis is a set of statistical processes for estimating the relationships between a dependent variable (often called the 'outcome variable') and one or more independent variables (often called 'predictors,' 'covariates,' or 'features').
13. **k-Means Clustering**: k-means clustering is a method of vector quantization, originally from signal processing, that aims to partition *n* observations into k clusters in which each observation belongs to the cluster with the nearest mean (cluster centers or cluster centroid), serving as a prototype of the cluster.
14. **Expectation Maximization**: An expectation-maximization (EM) algorithm is an iterative method to find (local) maximum likelihood or maximum a posteriori (MAP) estimates of parameters in statistical models, where the model depends on unobserved latent variables.
15. **Hierarchical Clustering**: Hierarchical clustering (also called hierarchical cluster analysis or HCA) is a method of cluster analysis which seeks to build a hierarchy of clusters.
16. **Decision Tree**: Decision tree learning is one of the predictive modeling approaches used in statistics, data mining, and machine learning. It uses a decision tree (as a predictive model) to go from observations about an item (represented in the branches) to conclusions about the item's target value (represented in the leaves).
17. **k-NN (k-Nearest Neighbors)**: k-nearest neighbors algorithm (k-NN) is a non-parametric machine learning method used for classification and regression.
18. **Artificial Neural Network (ANN)**: Artificial neural networks (ANNs), usually simply called neural networks (NNs), are computing systems vaguely inspired by the biological neural networks that constitute animal brains.
19. **Multilayer Perception (MLP)**: A multilayer perceptron (MLP) is a class of feedforward artificial neural network (ANN). The term MLP is used ambiguously, sometimes loosely to any feedforward ANN, sometimes strictly to refer to networks composed of multiple layers of perceptron (with threshold activation);
20. **Random Forest**: Random forests or random decision forests are an ensemble learning method for classification, regression and other tasks that operate by constructing a multitude of decision trees at training time and outputting the class that is the mode of the classes (classification) or mean/average prediction (regression) of the individual trees.
21. **Support Vector Machines (SVMs)**: Support vector machines (SVMs) are supervised learning models with associated learning algorithms that analyze data for classification and regression analysis.

References

1. He, B., Zhang, D., Liu, S., Liu, H., Han, D., Ni, L.M.: Profiling driver behavior for personalized insurance pricing and maximal profit. In: 2018 IEEE International Conference on Big Data (Big Data), pp. 1387–1396. IEEE (2018)
2. Undavia, J.N., Dolia, P., Patel, A.: Customized prediction model to predict post-graduation course for graduating students using decision tree classifier. Indian J. Sci. Technol. **9**(12), 1–7 (2016)
3. Cevolini, A., Esposito, E.: From pool to profile: social consequences of algorithmic prediction in insurance. Big Data Soc. **7**(2) (2020)
4. Pandya, S.D., Virparia, P.V.: Folksonomy-based information retrieval by generating tag cloud for electronic resources management industries and suggestive mechanism for tagging using data mining techniques. In: Web Usage Mining Techniques and Applications Across Industries, pp. 80–91. IGI Global (2017)
5. Vaidya, N., Sajja, P., Gor, D.: Evaluating learning effectiveness in collaborative learning environment by using multi-objective grey situation decision making theory. Int. J. Sci. Eng. Res. (IJSER) **6**(8), 41–45 (2015)
6. Vaidya, N.M., Patel, K.K.: Learner performance and preference meter for better career guidance and holistic growth. In: ICT Analysis and Applications, pp. 47–54. Springer, Berlin (2020)
7. Vaidya, N.M., Undavia, J.N., Patel, A.: Multiple criteria decision making to improve retention ratio of employees by identifying and analyzing critical prevention factor. In: Information and Communication Technology for Competitive Strategies (ICTCS 2020), pp. 943–951. Springer, Berlin (2022)

Web Service Anti-patterns Detection Using CNN with Varying Sequence Padding Size

Sahithi Tummalapalli, Lov Kumar, and Neti Lalita Bhanu Murthy

1 Introduction

A web service is a collection of protocols and specifications that enable apps to communicate with one another. Web services have progressed as a result of the adoption of standards that facilitate interoperability. Distributed systems built on a services-oriented design may benefit from their use. When a small number of web services are combined, they may be used to build sophisticated distributed systems. A web service provider publishes certain functionality. Software developers can discover and call all essential web services to construct an application. There are several benefits of using web services; for example, they use SOAP mechanism which is more efficient as compared to regular HTTP. They are helpful in developing applications that are independent of programming languages.

Service-oriented architecture (SOA) enables developing different sorts of service-based systems (SBSs) similar to Amazon, eBay, Drop Box and many more, and the improvement of such systems raises numerous demanding issues. SBSs should evolve to match new user requirements and adapt new execution settings inclusive of incorporation of latest gadgets and technology. Due to a huge number of these alterations, SBSs' design and Quality of Service (QoS) may also be impaired. Anti-patterns, as they are known, typically result in a common negative remedy for recurring issues. These are the parts of the design that violate basic design principles and so have a

S. Tummalapalli (✉) · L. Kumar · N. L. Bhanu Murthy
Department of Computer Science and Information Systems, Birla Institute of Science and Technology-Pilani, Hyderabad Campus, Jawahar Nagar, Hyderabad, Telangana, India
e-mail: P20170433@hyderabad.bits-pilani.ac.in

L. Kumar
e-mail: lovkumar@hyderabad.bits-pilani.ac.in

N. L. Bhanu Murthy
e-mail: bhanu@hyderabad.bits-pilani.ac.in

J. Singh et al. (eds.), *Mobile Application Development: Practice and Experience*,
Studies in Systems, Decision and Control 452,
https://doi.org/10.1007/978-981-19-6893-8_13

detrimental impact on the final product's quality. There are many anti-patterns that are not accidental, and they are usually always followed with good intentions. While anti-patterns may make it difficult for a software system to evolve and improve, they can also be useful in spotting issues in the code, design and management of software projects, as well. The web service anti-patterns which we considered in this paper are GOWS: God Object Web Service (AP1), FGWS: Fine Grained Web Service (AP2), CWS: Chatty Web Service (AP3) and DWS: Data Web Service (AP4). Previous research showed that the presence of anti-patterns effect the performance and evolution of the software system [1]. Regardless of the broad use of Web services, no particular and automatic methodology for the detection of such anti-patterns from their Web Service Definition Language (WSDL) files exist to date. Ouni et al. [2] mentioned that automatic algorithms to detect and rectify these web services have been relatively unexplored and still in their infancy stage. Ouni et al. [2] was the first person to discover the approach in the year 2015. Similarly, Palma et al. [3] also highlighted the need to build automated approaches for the detection of web service anti-patterns.

The motivation behind the paper is thus to explore the techniques to detect the anti-patterns automatically using WSDL metrics. In this paper, we empirically investigate the performance of convolutional neural network (CNN) applied with word embedding technique on web service description language (WSDL) file in the detection of web service anti-patterns.

2 Objectives and Research Questions

The motivation of the work presented in this paper is to investigate the application of convolutional neural network (CNN) applied with word embedding technique in the prediction of SOA anti-patterns. The following research questions (RQ) have been answered in this work:

- **RQ1**: Is there a really essential differentiation between the performance displayed by the two data sampling techniques over the original data?
- **RQ2**: Is there a quantifiably enormous distinction between the performance of the models developed by utilizing word embedding technique with varying sequence padding sizes?
- **RQ3**: What is the general execution of CNN with distinct hidden layers considered with respect to AUC and F-measure metrics?

3 Related Work

Jaffar et al. [4] argued that classes engaging in anti-pattern and patterns of software designs have dependencies with other classes, i.e. unvarying and modifying dependencies, which may cause troubles to spread to other classes. Researchers in this study have empirically explored the implications of dependencies in object-oriented systems by highlighting and assessing the relationship between the presence of co-change and static dependencies and the change proneness, fault proneness and fault types of the classes. There is a method established by Velioğlu et al. [5] that can discover and minimize anti-patterns in the software project called Y-CSD. Specifically, the recommended technology is applied to find two anti-patterns: brain method and data class. Y-CSD identifies code smells and anti-patterns using structural analysis. Code smell and anti-patterns may be decreased using the tool, Y-CSD, which minimizes the cost of software maintenance and helps new engineers adapt to current projects. Code smells and anti-patterns, which may lead to more significant concerns in future, are the subject of this study. Static code analysis by Kumar et al. [6] has been suggested as a technique for automatically discovering anti-patterns. An anti-pattern detection approach presented in this study focuses on the aggregate values of source code metrics generated at the web service level. An empirical study of eight machine learning algorithms (bagging, multilayer perceptrons, random forest, naive Bayes, decision tree, logistic boost, Adaboost, logistic regression), four data sampling techniques (downsampling, random sampling and synthetic minority oversampling technique (SMOTE) and four feature selection techniques (information gain), was conducted in this paper. Saluja et al. [7] proposed an unique optimized approach that uses both dynamic and static measures for execution. Genetic algorithms are applied to enhance the results. The novel techniques yielded better results than the present methods, with a recall rate of about 0.9 for the suggested methods.

4 Experimental Setup

This section describes each module used in the proposed method, i.e. dataset, data sampling techniques, word embedding technique with CNN.

4.1 Experimental Dataset

In this work, the database we have used comprises 226 publicly available web services that are downloaded from the Tera-Promise repository in GitHub.[1] The dataset consists of a list of web services collected from various domains such as finance, weather, education, tourism and the anti-patterns present in them. The statistics and distribution of anti-patterns by type are shown in Fig. 1 [8, 9].

[1] https://github.com/ouniali/WSantipatterns.

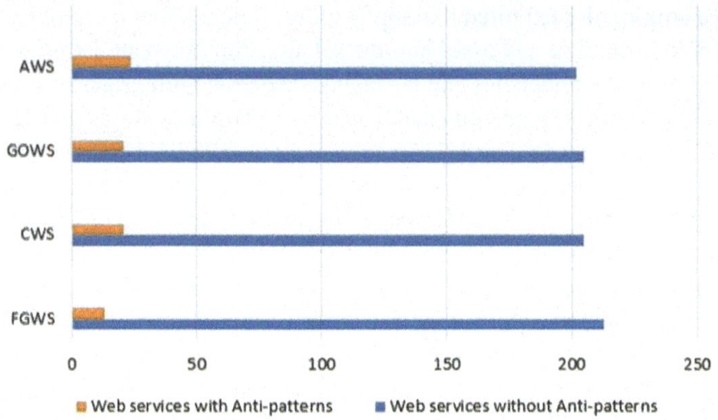

Fig. 1 Statistics on anti-patterns distribution by type

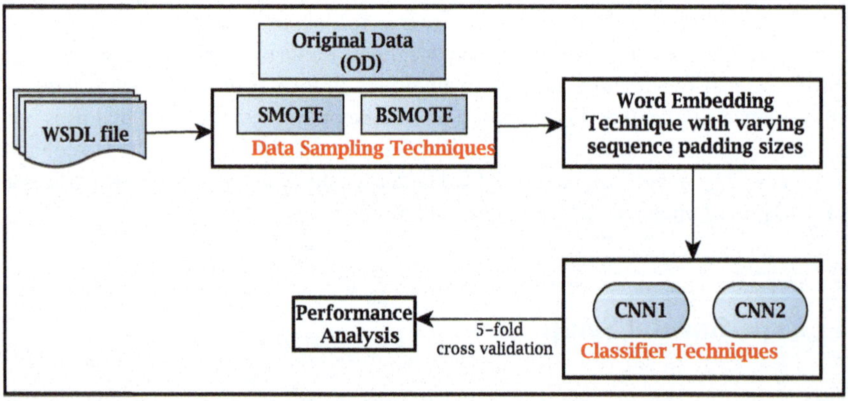

Fig. 2 Flowchart of proposed research framework

5 Proposed Methodology

Figure 2 shows the detailed illustration of the proposed research framework. As discussed in Sect. 4.1, we conducted our experiments on 226 publicly available web services that are downloaded from GitHub. We used two data sampling techniques SMOTE, Borderline SMOTE (BSMOTE), as stated in earlier sections, to eliminate the class imbalance problem. We used these two additional data sets and the original dataset (ORG) to train the models. Next, we applied word embedding technique with varying sequence padding sizes for feature generation. We then trained the models developed using CNN with hidden layers 1 and 2. Lastly, we applied fivefold cross-validation to repress the selection bias and overfitting problems. We then used performance metrics such as the area under the curve (AUC), F-measure, etc., and some tests like statistical significance testing to compare the performance of the various models developed.

6 Experimental Outcome

Table 1 shows accuracy for all the CNN models using feature generation techniques and data sampling techniques. Table 2, 4 and 6 shows the descriptive statistics of different metrics used in our work. From all the above tables stated, the following observations were inferred (Tables 3 and 5):

- Word embedding technique with sequence padding size as 200 performs better with a mean accuracy of 97.76%.
- SMOTE performs best among the data sampling techniques with a mean accuracy of 98.07% and a median accuracy of 98.39. Model developed using original data (OD) has the worst performance with a mean accuracy of 96.09%.
- CNN with two hidden layers is performing better with a mean accuracy of 97.58%.

6.1 Data Sampling Techniques

We can infer from Table 1 that there a bias between the web services which have anti-patterns and the web services that do not have anti-patterns. This is commonly referred to as the class imbalance problem. In order to mitigate the bias between the classes, we used two sampling techniques in this work to generate additional data sets. A short explanation of the sampling techniques is given below:

- **SMOTE** [10]: It uses the nearest neighbours of the minority class to generate the new samples artificially.
- **Borderline Smote (BSMOTE)** [11]: BSMOTE creates new instances of the minority class by employing the cases in the border area between classes that are closest neighbours to the cases in the minority class's closest neighbours.

6.2 Word Embedding with CNN

In this paper, we used word embedding technique to encode code in WSDL file as vector values. Each word of the WSDL file is mapped to a 32 size vector. The length of the sequence of words in a sentence may vary. Hence, we are developing models for different sequence padding sizes varying from 100 to 1000. For example, if we are developing model for padding size as 100, we will restrict each code line in WSDL file to be 100 words, diminishing lengthy code lines and padding short code lines with zero values.

Further, We use convolutional neural network [12] for training the model. Convolutional neural network consists of a couple of building blocks, consisting of convolution layers, pooling layers and completely linked layers and is designed to mechanically and adaptively research spatial hierarchies of features through a back-propagation algorithm. In this paper, we used CNN with two hidden layers, i.e. CNN with one hidden layer (CNN1) and CNN with two hidden layers (CNN2) for training

Table 1 Accuracy: all models

Sampling technique	Anti-pattern	Sequence padding length	CNN1	CNN2	Sampling technique	Anti-pattern	Sequence padding length	CNN1	CNN2
OD	AP1	100	0.93	0.91	OD	AP3	100	0.96	0.96
OD	AP1	200	0.94	0.95	OD	AP3	200	0.97	0.98
OD	AP1	300	0.94	0.96	OD	AP3	300	0.97	0.97
OD	AP1	400	0.95	0.96	OD	AP3	400	0.96	0.98
OD	AP1	500	0.94	0.95	OD	AP3	500	0.97	0.97
OD	AP1	600	0.96	0.96	OD	AP3	600	0.97	0.97
OD	AP1	700	0.95	0.97	OD	AP3	700	0.96	0.96
OD	AP1	800	0.92	0.95	OD	AP3	800	0.98	0.97
OD	AP1	900	0.94	0.95	OD	AP3	900	0.97	0.98
OD	AP1	1000	0.94	0.95	OD	AP3	1000	0.97	0.97
SMOTE	AP1	100	0.97	0.95	SMOTE	AP3	100	0.98	0.98
SMOTE	AP1	200	0.98	0.98	SMOTE	AP3	200	0.99	0.99
SMOTE	AP1	300	0.98	0.97	SMOTE	AP3	300	0.99	0.99
SMOTE	AP1	400	0.98	0.97	SMOTE	AP3	400	0.99	0.99
SMOTE	AP1	500	0.96	0.96	SMOTE	AP3	500	0.99	0.99
SMOTE	AP1	600	0.96	0.96	SMOTE	AP3	600	0.99	0.98
SMOTE	AP1	700	0.98	0.99	SMOTE	AP3	700	0.99	0.99
SMOTE	AP1	800	0.97	0.98	SMOTE	AP3	800	0.99	0.99
SMOTE	AP1	900	0.97	0.97	SMOTE	AP3	900	0.98	0.98
SMOTE	AP1	1000	0.97	0.97	SMOTE	AP3	1000	0.99	0.98
BLSMOTE	AP1	100	0.96	0.96	BLSMOTE	AP3	100	0.98	0.98
BLSMOTE	AP1	200	0.96	0.96	BLSMOTE	AP3	200	1.00	0.99
BLSMOTE	AP1	300	0.97	0.97	BLSMOTE	AP3	300	0.99	0.99
BLSMOTE	AP1	400	0.98	0.98	BLSMOTE	AP3	400	0.99	0.99
BLSMOTE	AP1	500	0.98	0.99	BLSMOTE	AP3	500	0.99	0.99
BLSMOTE	AP1	600	0.97	0.95	BLSMOTE	AP3	600	0.91	0.98
BLSMOTE	AP1	700	0.97	0.97	BLSMOTE	AP3	700	0.99	0.99
BLSMOTE	AP1	800	0.95	0.96	BLSMOTE	AP3	800	0.99	0.99
BLSMOTE	AP1	900	0.97	0.97	BLSMOTE	AP3	900	0.98	0.99
BLSMOTE	AP1	1000	0.97	0.96	BLSMOTE	AP3	1000	0.99	0.99
OD	AP2	100	0.97	0.97	OD	AP4	100	0.94	0.94
OD	AP2	200	0.97	0.98	OD	AP4	200	0.94	0.96
OD	AP2	300	0.96	0.97	OD	AP4	300	0.96	0.96
OD	AP2	400	0.96	0.98	OD	AP4	400	0.94	0.94
OD	AP2	500	0.96	0.97	OD	AP4	500	0.97	0.96
OD	AP2	600	0.96	0.98	OD	AP4	600	0.94	0.95
OD	AP2	700	0.96	0.99	OD	AP4	700	0.94	0.97
OD	AP2	800	0.96	0.98	OD	AP4	800	0.95	0.94
OD	AP2	900	0.98	0.99	OD	AP4	900	0.96	0.96
OD	AP2	1000	0.95	0.94	OD	AP4	1000	0.93	0.95
SMOTE	AP2	100	0.99	0.99	SMOTE	AP4	100	0.93	0.96
SMOTE	AP2	200	0.99	0.99	SMOTE	AP4	200	0.99	0.98
SMOTE	AP2	300	0.99	0.98	SMOTE	AP4	300	0.98	0.98

(continued)

Table 1 (continued)

Sampling technique	Anti-pattern	Sequence padding length	CNN1	CNN2	Sampling technique	Anti-pattern	Sequence padding length	CNN1	CNN2
SMOTE	AP2	400	1.00	1.00	SMOTE	AP4	400	0.96	0.99
SMOTE	AP2	500	0.98	0.98	SMOTE	AP4	500	0.97	0.99
SMOTE	AP2	600	0.97	0.98	SMOTE	AP4	600	0.99	0.99
SMOTE	AP2	700	0.98	0.97	SMOTE	AP4	700	0.99	0.99
SMOTE	AP2	800	0.98	0.98	SMOTE	AP4	800	0.98	0.96
SMOTE	AP2	900	0.99	0.99	SMOTE	AP4	900	0.97	0.98
SMOTE	AP2	1000	0.99	0.99	SMOTE	AP4	1000	0.97	0.97
BLSMOTE	AP2	100	0.98	0.97	BLSMOTE	AP4	100	0.97	0.97
BLSMOTE	AP2	200	0.99	0.99	BLSMOTE	AP4	200	0.98	0.98
BLSMOTE	AP2	300	0.99	0.99	BLSMOTE	AP4	300	0.98	0.99
BLSMOTE	AP2	400	0.98	0.98	BLSMOTE	AP4	400	0.98	0.98
BLSMOTE	AP2	500	0.99	0.99	BLSMOTE	AP4	500	0.99	0.98
BLSMOTE	AP2	600	0.97	0.99	BLSMOTE	AP4	600	0.98	0.99
BLSMOTE	AP2	700	1.00	0.99	BLSMOTE	AP4	700	0.95	0.98
BLSMOTE	AP2	800	0.94	0.98	BLSMOTE	AP4	800	0.99	0.99
BLSMOTE	AP2	900	0.99	0.99	BLSMOTE	AP4	900	0.97	0.97
BLSMOTE	AP2	1000	0.99	0.99	BLSMOTE	AP4	1000	0.98	0.98

Table 2 Descriptive statistics: sampling techniques

	Min	Max	Mean	Median	Var	Q1	Q3
Accuracy							
OD	91.41	98.99	96.09	96.46	2.54	94.95	97.47
SMOTE	93.33	100.00	98.07	98.39	1.40	97.24	98.92
BLSMOTE	90.98	99.73	97.94	98.33	2.02	97.14	98.91
F-mean							
OD	0.15	0.89	0.64	0.67	0.03	0.57	0.74
SMOTE	0.93	1.00	0.98	0.98	0.00	0.97	0.99
BLSMOTE	0.92	1.00	0.98	0.98	0.00	0.97	0.99
AUC							
OD	0.52	1.00	0.93	0.95	0.01	0.89	0.99
SMOTE	0.98	1.00	0.99	1.00	0.00	0.99	1.00
BLSMOTE	0.97	1.00	0.99	1.00	0.00	0.99	1.00

Table 3 Rank sum test: sampling techniques

	OD	SMOTE	BLSMOTE
OD	1	8.57E−12	4.84E−13
SMOTE	8.57E−12	1	0.521146
BLSMOTE	4.84E−13	0.521146	1

Table 4 Descriptive statistics: sequence padding sizes

Sequence padding size	Min	Max	Mean	Median	Var	Q1	Q3
Accuracy							
100	91.41	98.67	96.45	96.81	3.73	95.49	98.24
200	93.94	99.73	97.76	98.19	2.50	96.96	98.80
300	93.94	99.47	97.72	97.78	1.92	96.97	98.79
400	93.94	100.00	97.68	98.12	2.79	96.46	98.93
500	94.44	99.19	97.57	97.98	1.79	96.55	98.65
600	90.98	99.19	97.03	97.15	3.59	96.05	98.44
700	93.94	99.73	97.70	97.65	2.35	96.60	98.96
800	92.42	99.44	97.03	97.56	3.40	95.75	98.56
900	93.94	99.20	97.58	97.64	1.75	96.95	98.57
1000	92.93	99.47	97.16	97.24	3.40	95.83	98.63
F-mean							
100	0.45	0.99	0.86	0.96	0.03	0.67	0.98
200	0.50	1.00	0.89	0.98	0.02	0.77	0.99
300	0.45	0.99	0.87	0.98	0.03	0.73	0.99
400	0.33	1.00	0.87	0.98	0.03	0.79	0.99
500	0.33	0.99	0.87	0.98	0.03	0.73	0.99
600	0.33	0.99	0.86	0.97	0.03	0.74	0.98
700	0.46	1.00	0.88	0.97	0.03	0.83	0.99
800	0.21	0.99	0.86	0.97	0.04	0.75	0.98
900	0.45	0.99	0.90	0.97	0.02	0.82	0.98
1000	0.15	0.99	0.82	0.97	0.07	0.64	0.99
AUC							
100	0.76	1.00	0.96	0.99	0.00	0.97	0.99
200	0.85	1.00	0.98	0.99	0.00	0.99	1.00
300	0.89	1.00	0.97	0.99	0.00	0.96	1.00
400	0.83	1.00	0.97	1.00	0.00	0.97	1.00
500	0.79	1.00	0.97	0.99	0.00	0.99	1.00
600	0.87	1.00	0.98	0.99	0.00	0.98	1.00
700	0.81	1.00	0.98	0.99	0.00	0.99	1.00
800	0.81	1.00	0.97	0.99	0.00	0.98	1.00
900	0.87	1.00	0.98	1.00	0.00	0.99	1.00
1000	0.52	1.00	0.95	0.99	0.01	0.97	1.00

Table 5 Rank sum test: sequence padding sizes

	100	200	300	400	500	600	700	800	900	1000
100	1.00	0.03	0.24	0.02	0.01	0.18	0.02	0.11	0.00	0.29
200	0.03	1.00	0.51	0.47	0.99	0.63	0.68	0.57	0.29	0.57
300	0.24	0.51	1.00	0.35	0.39	0.84	0.18	0.89	0.10	0.80
400	0.02	0.47	0.35	1.00	0.66	0.33	0.92	0.27	0.78	0.42
500	0.01	0.99	0.39	0.66	1.00	0.43	0.83	0.47	0.56	0.43
600	0.18	0.63	0.84	0.33	0.43	1.00	0.22	0.93	0.13	0.90
700	0.02	0.68	0.18	0.92	0.83	0.22	1.00	0.37	0.87	0.23
800	0.11	0.57	0.89	0.27	0.47	0.93	0.37	1.00	0.16	0.88
900	0.00	0.29	0.10	0.78	0.56	0.13	0.87	0.16	1.00	0.19
1000	0.29	0.57	0.80	0.42	0.43	0.90	0.23	0.88	0.19	1.00

Table 6 Descriptive statistics: CNN

	Min	Max	Mean	Median	Var	Q1	Q3
Accuracy							
CNN1	90.98	99.73	97.16	97.40	3.27	96.29	98.63
CNN2	91.41	100.00	97.58	98.06	2.25	96.47	98.66
F-mean							
CNN1	0.18	1.00	0.84	0.97	0.05	0.67	0.99
CNN2	0.15	1.00	0.89	0.97	0.02	0.78	0.99
AUC							
CNN1	0.52	1.00	0.97	0.99	0.00	0.99	1.00
CNN2	0.73	1.00	0.97	0.99	0.00	0.99	1.00

the anti-pattern prediction models. For CNN1, the first layer is the embedded layer that takes 32 size vectors (represents words in WSDL file) as input. Then, we added CNN and max pooling layers after the embedding layer. The last layer is single neuron with sigmoid activation function, as we are dealing with classification problem with two classes: not anti-pattern (0) and anti-pattern (1).

7 Competitive Analysis

RQ1: Is there a really essential differentiation between the performance displayed by the two data sampling techniques over the original data?
Table 2 and Fig. 3 shows that the SMOTE is performing better across the AUC and F-measure metrics. SMOTE performs better than BLSMOTE, as SMOTE uses all of minority instances in the complete location to generate synthetic data. Compared to the accuracy measure of all the models, the model developed using the original dataset (OD) shows worst performance.

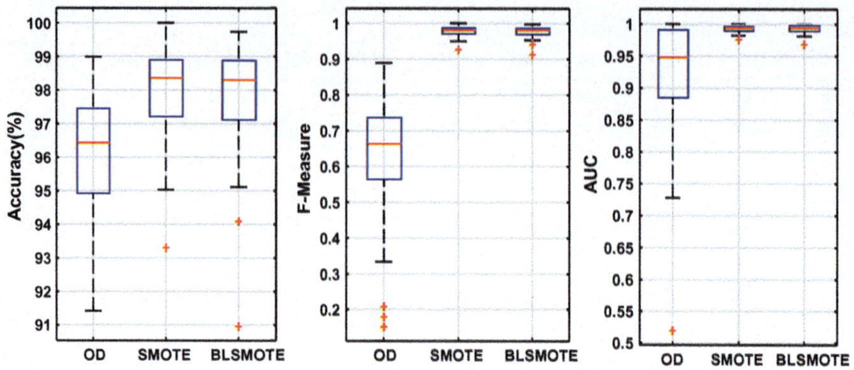

Fig. 3 Box-plot for accuracy, F-measure and AUC: data sampling techniques

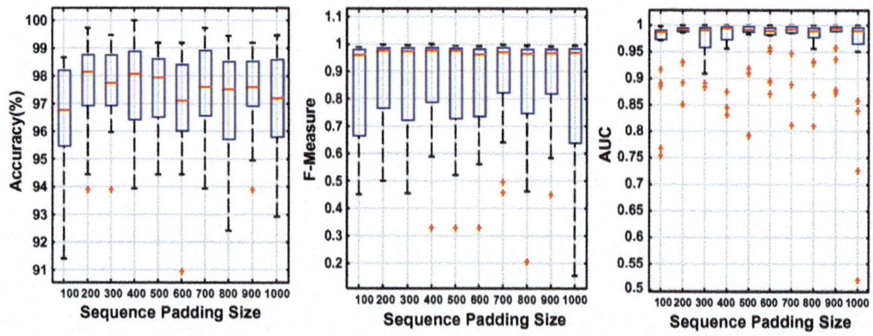

Fig. 4 Box-plot for accuracy, F-measure and AUC: sequence padding size

Table 3 gives the results of the rank sum test of the datasets generated using SMOTE, BLSMOTE and OD. We inferred that the models' performance trained using sampling techniques datasets varies significantly from the original dataset. We conclude that the performance values of all the datasets are highly uncorrelated. We also observed that the performance values of the models developed using SMOTE and BLSMOTE are similar.

RQ2: Is there a quantifiably enormous distinction between the performance of the models developed by utilizing word embedding technique with varying sequence padding sizes?
Table 4 and Fig. 4 show that the embedding technique with sequence padding size as 200 performs better than other models with varying padding size. Our model performs better with a sequence length of 200, indicating that the size of most of the sentences given as input is around 200.

In this research, the Wilcoxon signed-rank test is used to compare the prediction effectiveness of web service anti-pattern detection approaches utilizing varied padding sizes. The fundamental reason for doing this statistical testing is to deter-

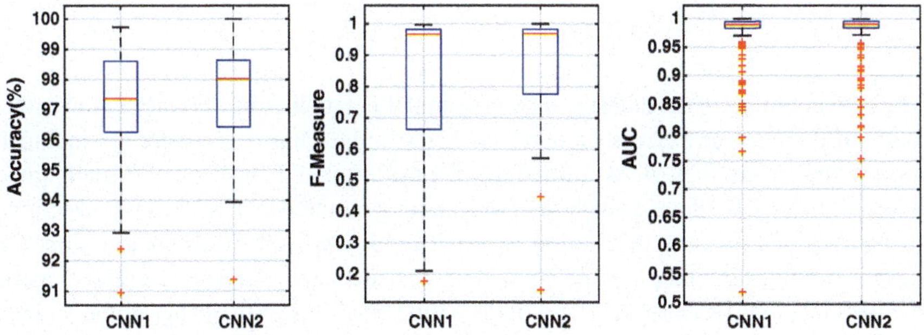

Fig. 5 Box-plot for accuracy, F-measure and AUC: CNN

Table 7 Rank sum test: CNN

	CNN1	CNN2
CNN1	1.00	0.96
CNN2	0.96	1.00

mine whether or not the models constructed using various sequence padding sizes have a statistically meaningful improvement. In order to evaluate whether or not to accept or reject the null hypothesis, *p*-value is utilized in this test. "The web service anti-pattern detection models constructed using varied sequence padding sizes are substantially the same," is the investigated null hypothesis for this study. Null hypothesis is accepted when rank sum test results are larger than 0.05 *p*-value. Wilcoxon signed-rank sum tests of vectors produced for all models with padding sizes ranging from 100 to 1000 are shown in Table 5. When we examined Table 5, we found the majority of comparison points had values larger than 0.05 by comparison. It follows from this that, in most circumstances, we can infer that the models built by taking into account various sequence padding lengths as input are considerably different from each other. Thus, the performance of the models produced may be determined by the duration of the padding sequences.

RQ3: What is the general execution of CNN with distinct hidden layers considered with respect to AUC and f-measure metrics?
Figure 5 and Table 6 show the performance of two classifier techniques developed in terms of accuracy, AUC and F-measure. From Fig. 5 and Table 6, we infer that the CNN2 is performing slightly better when compared to other models with a mean accuracy of 97.58%. This might be because the CNN model with two hidden layers detects more complex features while training, while CNN model with 1 hidden layer might have failed to detect more complex features. In Table 7, the rank sum tests on CNN models with different number of hidden layers are shown. Table 7 shows that the prediction models generated using various CNN models are notably distinct from each other and very unrelated.

8 Threats to Validity

The dependability of the dataset is one potential threat to internal validity found in this study. If there are any errors or inconsistencies in the dataset that were not included in the experiment, they were omitted from consideration. Consistency is maintained when collecting the data, even though we cannot guarantee that the dataset is accurate to the tenth of a per cent. In addition, external validity is a potential concern, since the framework does not take into account elements such as the expertise of developers, the history of system progress, the underlying principles of the programme and the many types of developers and stakeholders involved.

9 Conclusion and Future Scope

Various CNN models, word embedding approaches with different sequence padding widths and data sampling strategies are all used in this work to offer an empirical assessment of anti-pattern prediction. In order to compare the models accuracy, AUC and F-measures, we used a fivefold cross-validation procedure. The following are the key outcomes of our research:

- CNN with two hidden layers is performing better with a mean accuracy of 97.58%.
- SMOTE performs best among the data sampling techniques with a mean accuracy of 98.07% and a median accuracy of 98.39. Model developed using original data (OD) has the worst performance with a mean accuracy of 96.09%.
- Word embedding technique with sequence padding size as 200 performs better with a mean accuracy of 97.76%.

NLP approaches will be used in future to construct more accurate models for predicting web service anti-patterns. Finding out the performance of web service anti-pattern prediction models based on various feature selection strategies, data sampling techniques and classifier techniques on the metric set produced by performing diverse NLP techniques on the WSDL files would be interesting.

References

1. Tummalapalli, S., Bhanu Murthy, N.L., Krishna, A., et al.: Detection of web service anti-patterns using neural networks with multiple layers. In: International Conference on Neural Information Processing, pp. 571–579. Springer (2020)
2. Ouni, A., Kessentini, M., Inoue, K., Cinnéide, M.O.: Search-based web service antipatterns detection. IEEE Trans. Serv. Comput. **10**(4), 603–617 (2015)
3. Palma, F., Nayrolles, M., Moha, N., Guéhéneuc, Y.-G., Baudry, B., Jézéquel, J.-M.: SOA antipatterns: an approach for their specification and detection. Int. J. Cooper. Inf. Syst. **22**(04), 1341004 (2013)

4. Jaafar, F., Guéhéneuc, Y.-G., Hamel, S., Khomh, F., Zulkernine, M.: Evaluating the impact of design pattern and anti-pattern dependencies on changes and faults. Empir. Softw. Eng. **21**(3), 896–931 (2016)
5. Velioğlu, S., Selçuk, Y.E.: An automated code smell and anti-pattern detection approach. In: 2017 IEEE 15th International Conference on Software Engineering Research, Management and Applications (SERA), pp. 271–275. IEEE (2017)
6. Kumar, L., Sureka, A.: An empirical analysis on web service anti-pattern detection using a machine learning framework. In: 2018 IEEE 42nd Annual Computer Software and Applications Conference (COMPSAC), vol. 1, pp. 2–11. IEEE (2018)
7. Saluja, S., Batra, U.: Optimized approach for antipattern detection in service computing architecture. J. Inf. Optim. Sci. **40**(5), 1069–1080 (2019)
8. Tummalapalli, S., Kumar, L., Lalita Bhanu Murthy, N.: An empirical framework for web service anti-pattern prediction using machine learning techniques. In: 2019 9th Annual Information Technology, Electromechanical Engineering and Microelectronics Conference (IEMECON), pp. 137–143. IEEE (2019)
9. Tummalapalli, S., Kumar, L., Neti, L.B., Krishna, A.: Detection of web service anti-patterns using weighted extreme learning machine. Comput. Stand. Interfaces 103621 (2022)
10. Chawla, N.V., Bowyer, K.W., Hall, L.O., Kegelmeyer, W.P.: SMOTE: synthetic minority over-sampling technique. J. Artif. Intell. Res. **16**, 321–357 (2002)
11. Han, H., Wang, W.-Y., Mao, B.-H.: Borderline-smote: a new over-sampling method in imbalanced data sets learning. In: International Conference on Intelligent Computing, pp. 878–887. Springer (2005)
12. Kalchbrenner N, Grefenstette E, Blunsom P (2014) A convolutional neural network for modelling sentences. arXiv preprint arXiv:1404.2188

The manufacturer's authorised representative in the EU is Springer Nature Customer Service Centre GmbH, Europaplatz 3, 69115 Heidelberg, Germany. If you have any concerns regarding our products, please contact ProductSafety@springernature.com

Printed and bound by CPI Group (UK) Ltd, Croydon, CR0 4YY

30/09/2026

02225336-0003